REGIONAL INPUT-OUTPUT ANALYSIS

GEOFFREY J.D. HEWINGS
Department of Geography and Regional Science Program
University of Illinois at Urbana-Champaign

SAGE PUBLICATIONS
The Publishers of Professional Social Science
Beverly Hills London New Delhi

For information address:

SAGE Publications, Inc.
275 South Beverly Drive
Beverly Hills, California 90212

SAGE Publications India Pvt. Ltd.
M-32 Market
Greater Kailash I
New Delhi 110 048 India

SAGE Publications Ltd
28 Banner Street
London EC1Y 8QE
England

International Standard Book Number 0-8039-2740-1
International Standard Book Number 0-8039-2521-2 (pbk)

Library of Congress Catalog Card No. L.C. 85-14220

FIRST PRINTING

CONTENTS

ACKNOWLEDGMENTS

The comments and suggestions provided by Gordon Mulligan, Adrian Esparza, Grant Thrall, Randall Johnson, Breadan O'hUallachain, and the reviewer, Adam Rose, are gratefully appreciated. My introduction to regional input-output analysis was provided by Charles Tiebout almost twenty years ago. William Miernyk's *Elements of Input-Output Analysis* served as an excellent reference for the novitiate input-output analyst. Modern regional analysis owes a great deal to their important contributions; I hope the spirit of their influence is contained within this small volume.

SERIES EDITOR'S INTRODUCTION

Input-output analysis is a method by which the flow of production can be traced among the various sectors of the economy, through to final demand or export. The most fundamental problem of input-output analysis is to calculate the necessary output levels of each industry required to achieve a final output. Included among the uses of input-output analysis is the ability to determine the following: What is the effect upon the local economy from the introduction of a new firm? What are the economic linkages between regions and how is equilibrium between regions achieved? What if the supply of an input in one region becomes restricted through some bottleneck?

The foundations of modern input-output analysis can be traced to work in both economics and geography. The linkages to geography have been largely through the earlier analysis of urban economic base and city classification. In economics the work can be traced to the 1930s pioneering efforts of W. W. Leontief, which led to his receiving the Nobel Prize. Much of the recent contributions in input-output analysis falls within the purview of regional science, the overlap of interests in economics, geography, city and regional planning, and engineering. This literature has refined the ability of input-output analysts to work with incomplete data sets, arrive at stable and accurate estimates, and apply the general input-output method in practice to actual planning situations. With the exception of gravity and spatial interaction models, no topic in scientific geography has achieved greater practical application.

Professor Geoffrey Hewings is recognized as one of the leading contemporary scholars in input-output modeling, and has presented here one of the most readable introductions to the input-output problem and the contemporary literature that refines the technique. At the same time, save for the requirement that the reader have a grasp of elementary matrix algebra, Hewings keeps the book entirely at a level that can be understood by a reader who is encountering the material for the first time.

Professor Hewings first establishes the historical links between input-output models and the earlier macroeconomic accounting framework, economic base models, and the fundamental regional input-output model. Following the development of the basic model, and a discussion of the interpretation of the components of the model such as income and employment multipliers, Hewings discusses how the general model can be applied in practice. On applying the model he first describes how to construct the input-output tables for interregional and multiregional input-output matrices; he then presents a general discussion of estimation; and finally he presents practical examples of implementing the input-output model. The book brings the reader up to a discussion of the contemporary research frontier and likely future developments in input-output analysis.

This book will prove to be a valuable resource to students and practitioners of the planning sciences, including urban and regional economics, regional science, engineering, public administration, business management science, city and regional planning, as well as scientists in economic geography.

—*Grant Ian Thrall*
Series Editor

REGIONAL INPUT-OUTPUT ANALYSIS

GEOFFREY J.D. HEWINGS
Department of Geography and Regional Science Program
University of Illinois at Urbana-Champaign

1. INTRODUCTION

Imagine a national economy that has been divided into a set of regions. Within each region, grouped into sectors, there is a set of firms producing a variety of commodities that are consumed by other firms in the course of the production of other more finished commodities (e.g., automobile parts are assembled into a finished automobile), consumers, government, export markets, or other firms using these commodities as investment goods. In addition to engaging in sales activities, firms are also active in the purchase of commodities and other inputs—labor, entrepreneurial skills, as well as commodities purchased from outside the region. It would not be unusual to find in a region of several million people well over 100,000 firms producing as many as half a million or more commodities.

Suppose that a newly elected national-level or federal government proposes to reorient priorities away from social spending to defense spending. What will be the resulting impact on our regional economy and other regions making up this nation? As a result of these changes or in response to other stimuli, assume a new firm locates in the region employing 2000 people. What will be the impact of this new activity on the region? From another perspective, assume that the comparative advantage that the region once enjoyed in the export of its commodities is eroded, with the resulting closure of many local firms and an increase in regional unemployment levels. Again, what will be the impact of this activity change on the regional economy?

With the large number of firms, commodities, consumers, and other actors in the regional economy, it should be obvious that tracing the impacts on a firm-by-firm or consumer-by-consumer basis would be a daunting proposition. Clearly, we need some accounting system into which these interactions can be placed in the hope that some analytical method could be employed to trace the impacts in a systematic fashion. In a sense, we are going to have to sacrifice the richness of the reality of the regional economy for some reduced-form picture or model that is tractable and, we hope, representative as far as is possible of the micro level interactions. As happens with a great deal of analytical work in the social sciences, the gains from model development are not without cost; as we shall see, this is the case in the development of a family of analytical tools that are referred to as *social accounting systems.* Regional input-output analysis is one subset of these accounting systems.

This book will focus on some of the more elementary versions of these social accounting models (or SAMs) to provide a guide to their underlying theoretical structure and to explore ways in which they can be used to answer the sort of questions posed in the preceding paragraph. Thereafter, some excursions will be made into new developments that have extended the range and analytical complexity of these models, thereby enabling us to use them to answer more complex questions. One of the most interesting features of SAMs is that, on the one hand, they are strongly linked with standard macroeconomic accounting principles, and, on the other hand, they can be linked with many of the more traditional avenues of inquiry in the geographic and regional science fields. For example, interest in spatial interaction of commodities or individuals can be linked with a SAM framework; hence, we have the capability to explore the effects of the federal program changes alluded to earlier, not only on the structure of the economic system but potentially on the degree to which these changes will in turn promote changes in regional attractiveness for migration decisions.

These changes, in their turn, will have a further impact on the structure of production in the regional economy. How? Consider the case of a new firm opening up in the region. Because we are assuming a freely mobile society, with no restrictions on interregional movement, competition for the new jobs may come not only from local residents but from other persons inmigrating from other regions. Assuming that total employment rises and that some of this increase is associated with inmigration, it is likely that the demand for local services will rise. The new inmigrants and their families will demand commodities, public services such as schools, health care and so forth, and thus create the necessary conditions for an expansion of the economic base of the region. Whether this occurs will depend in part on the degree of excess capacity that may already exist in some firms and public services.

Without getting too embroiled in the details at this stage, we can begin to see a very strong link emerging between the structure of *production* and the structure of *consumption*. Changes in either component of the regional economy are likely to lead to changes in the other and, in turn, to further changes in the first component. Viewing the regional economic system in this fashion—as a broadly based system of interdependence—provides substantial insights into the functioning of regional economies. It will enable us to begin to understand why regional economies may or may not be responsive to changes that may take place at the national or even at the international level, why some regional economies are exhibiting characteristics of decline, why others are growing, and why still others seem to be relatively immune from the effects of major structural changes that have been observed in many Western economies over the last two decades. The modeling systems to be described here will not provide answers to all our questions; many of these models contain very restrictive assumptions, precluding their use in many contexts. Neither are these models to be considered theories of regional economic growth and development. For the most part, they are empirical models that, although resting on some theoretical assumptions, are not exclusively associated with any one paradigm. In fact, these models have one very interesting attribute—they have been used in centrally planned, socialist, free market, developed, and developing economies alike. This flexibility provides one of their attractions.

The field of regional social accounting in general and input-output analysis in particular has a rich legacy. Some of the more prominent economists of this century have worked in this area; four—Tinbergen in 1969, Kuznets in 1971, Leontief in 1973, and Stone in 1984—have been awarded the Nobel Prize in Economics for their work in developing many of the accounting frameworks that will be used in this book. At the regional and interregional level, input-output analysis has attracted the interests of many scholars, among them Isard (the founder of the field of Regional Science), Tiebout, Moses, Miernyk, and Miller. Reference will be made to the contributions of these individuals throughout the text.

In the next chapter, we shall attempt to resolve the problem of linking all the actors in our regional economy in a way that will enable us to perform some simple analytical experiments. We will see how the input-output model was derived and how it is linked with some well-known models used by economists, geographers, and regional scientists. Chapter 3 will develop the basic analytical framework and derive the system of equations that drives the input-output model; Chapter 4 will explore some basic applications with this simple model. The construction of regional input-output models will be addressed briefly in Chapter 5. Thereafter, in Chapter 6, we shall branch out to consider the ways in which this model can be expanded from a one-region version to consider interaction among two or more regions. It will be here that we find a link with other popular models in the geographic literature—namely, gravity and spatial interaction models. The seventh chapter will provide an introduction to the ways in which this simple model can be extended, especially ways in which it can be linked with other analytical frameworks—such as linear programming and demographic models—to provide a more sophisticated representation of reality. The final chapter provides examples of some of the ways in which these models can be made more flexible and explores new directions in research. A guide to further reading is provided at the end of this final chapter.

Because the models to be described here rely on representation in matrix form, the reader might find it useful to refresh his or her memory of simple matrix operations prior to reading the next chapter. However, no proofs are provided for the existence of solutions. The major focus is on the understanding of the model structure and its workings.

2. LINKS BETWEEN ECONOMIC BASE, KEYNESIAN, AND INPUT-OUTPUT MODELS

In the first chapter, reference was made to some of the important contributors to the field of regional input-output analysis; many of these individuals were trained as economists and thus were strongly influenced by the Keynesian view of the functioning of the economic system. Many of the geographers who became associated with the emerging field of regional analysis and regional science were exposed to the ideas contained in the so-called basic-nonbasic ratio and its role in understanding the functioning of city systems. The basic-nonbasic idea was eventually recast to become the economic base model, one of the major contributors to the explanation of differential urban and regional growth in national economies. Another important influence on this field was the work being undertaken in international trade theory—particularly the role of the foreign trade multiplier. These seemingly disparate views of aspects of the economy, at different spatial levels (international,

national, and urban or regional), are really very closely related. In this chapter, we will begin by showing how they are linked. It is important to understand that methods of regional economic analysis may be linked directly with economic analysis directed at the national and international levels and with other standard accounting frameworks. The reasons for this will become evident in Chapter 7 when some of these links will be explored in more detail.

Standard Macro-Level Accounting Framework

Imagine a simplified national economy whose basic accounting identity can be shown as:

$$Y = C + I + G + E \qquad [2.1]$$

where Y represents gross national income or product, C refers to consumption, I to investment, G to government expenditure, and E to net exports (i.e., exports minus imports). Let us simplify equation 2.1 by lumping the following terms together:

$$E' = I + G + E \qquad [2.2]$$

and define the following relationship between consumption and income:

$$C = cY \qquad [2.3]$$

where c is the macro level (aggregate) average propensity to consume. The coefficient provides an indication of the disposition of an average dollar of income on consumption activities. Combining equation 2.2 into equation 2.1 and substituting for C from equation 2.3, we have:

$$Y = cY + E'$$

which yields:

$$Y - cY = E'$$

$$Y(1 - c) = E'$$

$$Y = [1/(1 - c)]\, E'$$

$$= (1 - c)^{-1}\, E' \qquad [2.4]$$

Equation 2.4 provides a link between a set of what might be termed "exogenous" forces (E′) and gross national income; given that $0 \leq c \leq 1$ (i.e., consumption is always smaller than income—we assume that in the aggregate people have a tendency to save some of their income), the expression $(1 - c)^{-1}$ will always be greater than 1. Hence, exogenous changes in the economy will always create an impact on gross national income that is greater than the initial exogenous change. How does this work? The explanation may prove clearer if equation 2.4 is rewritten as:

$$Y = (1 + c + c^2 + c^3 + c^4 \ldots\ldots)E' \qquad [2.5]$$

This expression provides a power series or rounds of spending expansion of equation 2.4 and provides a more direct interpretation of the processes involved. Assume that the initial impact, say, for example, an increase in export sales, is \$1 million (E′); the first effect on the economy will be this amount itself, the multiplication of E′ and 1 in equation 2.5. The second effect will be the product of c and E′; this tells us how much of the impact will be spent on consumption. Part of these consumer expenditures will become income to other individuals who work in places offering consumer goods or in factories producing them or in sectors that are involved in wholesale, retail, or transportation of these goods to market; this portion is c^2E'. Again, part of these expenditures will become income and so the process continues. As c is less than 1, the proportion that becomes additional income at each round is smaller and smaller until, for all intents and purposes, it becomes zero. If we assume that c = 0.8, the various rounds of spending may be summarized in Figure 2.1. This process, whereby an initial change yields a final effect that is greater, is known as the *multiplier* process. In our example, with E′ = \$1m and c = 0.8, the effect on Y will be \$5m. The term $(1 - c)^{-1}$ is known as the *aggregate multiplier*. As we will see later, this process may operate at any spatial scale.

The Economic Base Model

At the urban or regional level, geographers were able to identify two types of activities. These activities were variously referred to as city forming (exogenous, basic, export) and city filling (endogenous, nonbasic, local). The distinction was made in an attempt to understand the *raison d'être* for cities. The first set of activities—city forming—were said to provide the reasons for the city's existence; a combination of locational factors, access to a local raw material, or just chance may

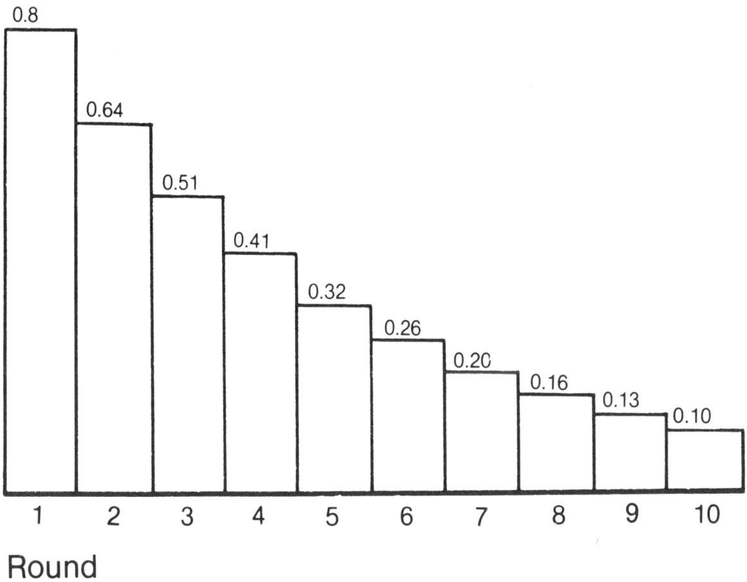

Figure 2.1 Rounds of Spending Multiplier Effect (Income per Round, Dollars)

have contributed to the activities locating there in the first place. Most certainly, industrial inertia (the tendency for firms to remain in their intitial locations) would have contributed to the continued presence of these activities in the city long after the initial forces may have dissipated. The city-forming activities produced goods and services for export outside the city or region; they were thus dependent upon markets over which they had little control. Serving the needs of these city-forming activities (through the provision of inputs) and the needs of the local population (retail goods and services) was a second set of activities, the city-filling functions. The latter existed only as a result of the presence of the city-forming functions, without whose presence the city would cease to exist. The number of city-filling activities was seen to be a function of the size of the city; here, we may see a link with another major explanatory model in geography, central place theory (see King 1984).

It was hypothesized that if the city-filling activities were dependent upon the city forming, then this link could be demonstrated quantitatively. Using more familiar terminology, the basic-nonbasic ratio was devised to this end. It was hypothesized that this ratio would change as a function of the size of the city; the larger the city, the greater the percentage of activities that would be classified as nonbasic. Assume that in a city with 100,000 employees, 40,000 worked in nonbasic and 60,000 in basic activities; we form the identity

$$E_t = E_{nb} + E_b \qquad [2.6]$$

where the subscripts refer to total, nonbasic, and basic employment, respectively. The basic-nonbasic ratio, r, may be defined as:

$$r = E_b / E_{nb} \qquad [2.7]$$

which, upon manipulation, reveals the relationship between nonbasic and basic employment:

$$E_{nb} = [1/r]E_b \qquad [2.8]$$

Using the above numerical example, we may see that for every three jobs created in the basic sector, an additional two jobs will be created in the nonbasic sector (because the expression [1/r] is 2/3). It remains now to use equation 2.8 to provide a relationship between basic and total activity by substitution in equation 2.7:

$$E_t = [1/r]\ E_b + E_b$$

$$E_t = [(1/r) + 1]E_b$$

$$= [(1 + r)/r]E_b \qquad [2.9]$$

In the numerical example, r = 3/2, hence the term in the square brackets is equal to 1.66; each job in basic activity will create 1.66 jobs in total (1 in basic and .66 in nonbasic). We may derive a similar expression to equation 2.9 in a way that is consistent with equation 2.4; instead of defining r as the basic-nonbasic ratio, define a as the proportion of total activity that is nonbasic:

$$a = E_{nb} / E_t \qquad [2.10]$$

This expression may be rewritten to yield:

$$E_{nb} = aE_t \quad [2.11]$$

which may now be substituted in equation 2.6 to provide the solution:

$$E_t = aE_t + E_b$$

$$E_t - aE_t = E_b$$

$$E_t(1 - a) = E_b$$

$$E_t = [1/(1 - a)]E_b$$

$$E_t = (1 - a)^{-1}E_b \quad [2.12]$$

Expression 2.12 thus provides the direct analogy to equation 2.4; the term $(1 - a)^{-1}$ may also be referred to as a multiplier—in this case, the economic base multiplier.[1]

The economic base model specifies a particular view of the economic growth and development process; namely, that it is generated by demand outside of the city or region. As such, it bears a strong relationship to the foreign trade multiplier in the international trade literature. There are, of course, a number of added complications and some very limiting assumptions with this formulation of the economy (see Hewings 1977; Isserman 1980; Gerking and Isserman 1981). In recent years, a number of important extensions to economic base analysis have been made. In particular, Ledent (1978), Gordon and Ledent (1981), Ledent and Gordon (1981), Mulligan and Gibson (1984), and Mathur and Rosen (1974) have made attempts to expand the scope of economic base analysis by considering new approaches to the identification of basic sectors, the links between economic and demographic change in a regional economy (see Chapter 7 for some attempts to explore this issue), and have attempted to expand our understanding of the role of nonwage and salary payments at the regional level.

The Link with the Input-Output Model

Although very rich in what they are able to accomplish, the models described above provide a very aggregate picture of a national and regional economy, respectively. The interactions taking place among the various actors in the system are not clearly specified; to do this, we will need to explore the economy in more detail. However, the linkages with the previous accounting systems will be maintained so that there will be a consistent format, enabling the analyst to move from one to the other with relative ease.

Romanoff (1974) was one of the first scholars to demonstrate a link between economic base and input-output analysis. If we maintain the strict assumptions of the economic base model, we may be able to recast the model in input-output terms. As we shall see later, a distinction is made in input-output analysis between purchases and sales made within the region and those made outside the region. In the economic base model, we will assume that (1) the basic sectors are the only ones that make sales outside the region, (2) the nonbasic sectors sell to each other and to the basic sector, and (3) there are no transactions among the basic sectors. In this case, two equations may now be specified; the subscripts 1 and 2 refer to nonbasic and basic sectors, respectively:

$$X_1 = T_{11} + T_{12} \quad [2.13]$$

$$X_2 = f_2 \quad [2.14]$$

where T_{11} and T_{12} are the flows among the nonbasic sectors and the flows from nonbasic to basic sectors, respectively, f_2 is the demand for the basic sectors' output outside the region, and X_1 and X_2 are the nonbasic and basic outputs, respectively. If we define the following proportions:

$$A_{11} = T_{11}/X_1 \quad [2.15]$$

$$A_{12} = T_{12}/X_2 \quad [2.16]$$

equation 2.13 may be rewritten by substituting T_{11} (= $A_{11}X_1$) and T_{12} (= $A_{12}X_2$) from equations 2.15 and 2.16, respectively:

$$X_1 = A_{11}X_1 + A_{12}X_2 \quad [2.17]$$

The proportions A_{11} and A_{12} may be thought of as the amount of nonbasic inputs needed to produce one unit of nonbasic output and the amount of nonbasic inputs needed in the production of one unit of basic output, respectively. The equations 2.17 and 2.14 may be shown in matrix form:

$$\begin{bmatrix} X_1 \\ \cdots \\ X_2 \end{bmatrix} = \begin{bmatrix} A_{11} & \vdots & A_{12} \\ \cdots & \vdots & \cdots \\ 0 & \vdots & 0 \end{bmatrix} \begin{bmatrix} X_1 \\ \cdots \\ X_2 \end{bmatrix} + \begin{bmatrix} 0 \\ \cdots \\ f_2 \end{bmatrix} \qquad [2.18]$$

This yields the solution for X_1:

$$X_1 = [I - A_{11}]^{-1} A_{12} f_2 \qquad [2.19]$$

because $f_2 = X_2$ (equation 2.14). This is the economic base model in an input-output formulation; the assumptions of the economic base system have been maintained. The result is a rather unusual system in which there is strict demarcation between the nonbasic and basic sectors of an economy. In reality, however, few sectors serve either the local or export market exclusively; hence we need to specify a more flexible version of equation 2.18 in order to accomplish this task.

The Regional Input-Output Accounting Framework: Initial Steps

Assume that we have the necessary funds to survey an appropriately stratified sample (by size and industrial composition) of all the firms within a region. From each firm, we would request information shown in Figure 2.2. The firm would be asked to detail its total purchases of goods (column 1) irrespective of the geographic origin of these purchases; similarly, the firm would be asked to specify the industrial sectors purchasing its output (column 3). In addition, the firms would be asked to provide information on purchases of labor and returns to capital (profits, dividends, and taxes) and sales to consumers, government, and for investment purchases. These transactions will be explained in greater detail later on in the next chapter. As we are dealing with a regional economy, we are going to be interested in the transactions that take place within the region as opposed to those with firms outside the region. For this reason, the firms in the survey would be asked to provide an estimate of their purchases (column 2) and sales (column 4) that occur within the region.

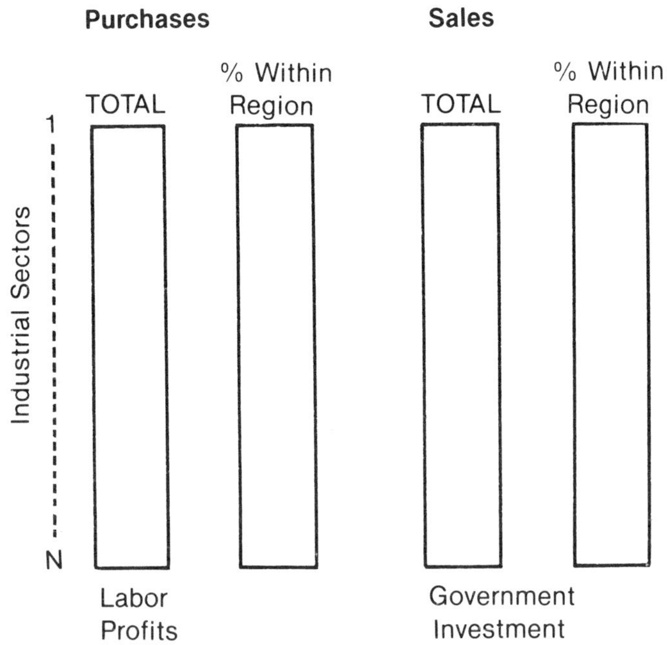

Figure 2.2 Survey Information Requested from Each Firm

There are a number of major, tricky accounting issues that have to be resolved in order to move from the sample data to an input-output table. These will be mentioned here but not explained in great detail (see Miernyk 1965). First, some rules have to be adopted to allocate firms to sectors. In the format that will be used in this book, the rule is usually based on the principal product of the firm.[2] The number of sectors identified will vary; some models show as few as 10, others as many as 400. Much will depend on the funds available for survey analysis and whether, in fact, survey data can be collected at all.[3] Furthermore, flows are usually shown in producers' rather than purchasers' prices; this necessitates allocating trade and transportation margins to the appropriate sectors. These margins may be thought of as the "mark-up" that a wholesaler or retailer charges for the service of marketing that they provide. In some cases, this may be a fixed percentage of the cost of the good (e.g., 25 percent). The transportation margins are an estimate of

(a) Transportation

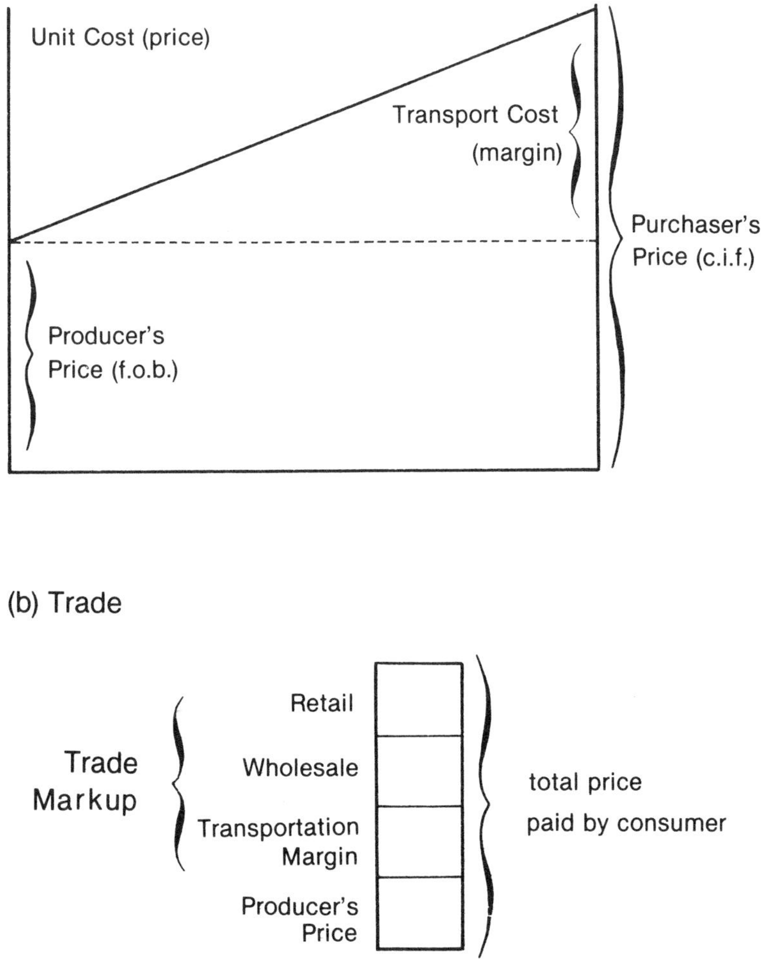

NOTE: In this example, the consumer purchase would involve four sectors (1) industrial sector producing good, (2) transport service, (3) wholesale, and (4) retail trade.

Figure 2.3 Trade and Transportation Margins

the transportation cost involved in moving a unit value of the commodity from producer to consumer. Hence, the purchase of a good by another sector may be shown as a combination of the purchase of the good and the purchase of the transport service in having the good delivered (see Figure 2.3a). For a purchase by a consumer, the transaction may involve both a wholesale and a retail mark-up (Figure 2.3b). That is why, in input-output models, consumers are shown as purchasing output from industries directly rather than aggregating all their transactions into the retail sector. A final issue revolves around the problem that as we are using a *sample* of firms, what all firms in sector i say they sell to sector j may not correspond to what the firms in sector j said they purchased from sector i. Hence, from the survey data, we end up with two matrices. The first documents the sales made by firms to all other sectors in the region whereas the second provides information on the purchases made by firms from other sectors. It would be very surprising if indeed the ij^{th} cell of both matrices contained the same value (suitably scaled to reflect the total population from which the sample was drawn). Accordingly, some careful rationalization of these two estimates must be made. This process has sparked a heated debate in the literature (see Gerking 1976, 1979a, 1979b; Miernyk 1976, 1979).

Now, we are ready to assemble the data into the regional input-output model. The format and the associated equation systems will form the subject of the next chapter. Keep in mind two important ideas; namely, the link with the economic base model and the relationship with national accounting systems of the Keynesian type.

3. THE BASIC REGIONAL INPUT-OUTPUT MODEL

In this chapter, we shall assemble the survey data obtained from our sample of firms into a simple regional input-output table. Once the assembly is complete, we may begin the process of using the table to develop a model for the purposes of multiplier estimation. In Chapter 4, some simple applications of the model and multipliers will be provided.

From the sample surveys, we now have assembled four matrices, one showing purchases made by firms from other firms irrespective of geographic origin, and a second one showing the purchases made from firms within the region. The third and fourth matrices display sales data irrespective of destination and within the region. The matrices showing the sales and purchases relations irrespective of origin and those showing those interactions within the region are "arbitraged" into two matrices. This arbitrage process ensures that consistency of estimation is obtained; it usually involves reconciliation of one or more estimates of the flows between two sectors. Because we are dealing with a system in which total inputs for all sectors equals total outputs for all sectors, any adjustments in the entries in one part of the matrix will require some adjustment to a number of other entries to ensure that balance is maintained. The first matrix (showing total transactions irrespective of origin) approximates what may be referred to as a total technology matrix, whereas the one that only details the transactions within the region is regarded as a regional transaction matrix. In the analysis that follows, most of our attention will be focused on the latter matrix. The transactions that involve purchases from outside the region are usually aggregated into one- or two-row vectors; a distinction is sometimes made between imports from other regions within the country and those from outside the country. The row vector or vectors are placed within the accounting system so that the extraregional transactions are not lost!

Elements of the Input-Output Table

Figure 3.1 shows the general accounting framework: the $n \times n$ square matrix shown in the double-lined box is known as the regional interindustry transactions matrix. A typical *row*, i, in this matrix shows the sales made by industry i to all other industries in the region; a typical column, j, shows the purchases made by this industry from all other industries in the region. What are these flows? They are sales and purchases made on *current* account and represent stages in the processing of *intermediate goods*. Current account purchases are those that a firm needs for the production of its commodities in any given year. Intermediate goods are those that are sold to other firms for further processing prior to sale to consumers. Examples would be sales of coal to an iron and steel plant (cell CS in Figure 3.1), sales of rolled steel to a fabricated metal plant (cell SF), sales of automobile bodies to an automobile assembly plant (cell BA), and so forth. However, the sale of a finished automobile to a consumer would not be shown in this part of the matrix. Where will this be found? To the right of the double-lined matrix is a box labeled "Final Demand." In this category are included the purchases by consumers, governments (local, state, and federal), and sales to other activities of investment goods.[4] The last category of final demand—exports—picks up the sales made by firms to purchasers outside the region and outside the country. Hence,

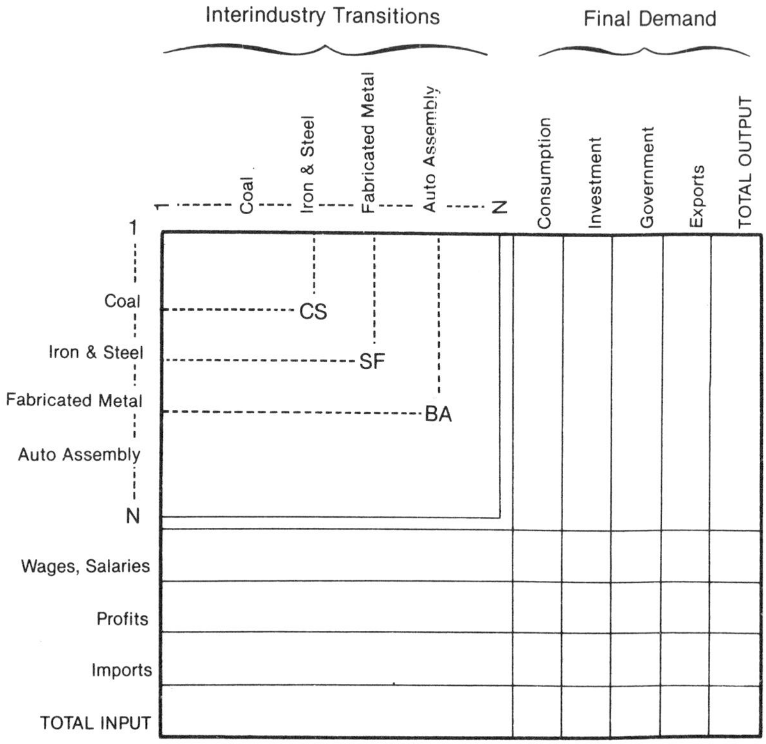

Figure 3.1 Schematic Input-Output Table

in parallel to the treatment of imports, two categories of export demand are shown in Figure 3.2—interregional and foreign. If we sum across the ith row, adding the sales to other industries and to the various categories of final demand, we arrive at the total *gross output* for that industrial sector.

Now examine the purchases made outside the square matrix in the jth column. The category known as value added contains two important elements: (1) returns to capital—such as profits and dividends; and (2) returns to labor, namely wages and salaries. Below that are the two import row vectors and finally, a vector of total *gross inputs*. One of the major contributions Leontief provided was the so-called "double-entry" accounting framework shown here; the vectors of total gross

From	1 Mining	2 Iron/ Steel	3 Electrical Engineering	4 Business Services	5 Trans-portation	6 Total Inter-mediate Sales	7 Households	8 State	9 Federal: Defense	10 Federal: Non-Defense	11 Inter-regional Exports	12 Foreign Exports	13 Total Final Demand	14 Total Output
1. Mining	21	–	9	3	–	33	30	10	5	–	20	2	67	100
2. Iron/steel	1	8	7	29	–	45	25	5	2	–	15	8	55	100
3. Electrical engineering	3	20	–	50	7	80	5	1	4	4	3	3	20	100
4. Business services	31	2	38	–	3	74	12	2	–	11	1	–	26	100
5. Trans-portation	10	25	26	1	4	66	9	6	–	13	4	2	34	100
6. Total inter-mediate purchases	66	55	80	83	14	298	81	24	11	28	–	–		
7. Value added	20	40	10	17	40	–	2	49	4	9	–	–		
8. Inter-regional imports	7	4	4	–	30	–	47	18	–	21	–	–		
9. Foreign imports	7	1	6	–	16	–	30	2	–	14	–	–		
10. Total input	100	100	100	100	100	–	160	93	15	72	–	–		

Figure 3.2 Regional Input-Output, 1984 (producers prices, $millions)

output and input are equal. Confronted with this fact, one might comment that if purchases and sales are equal, why are there firms in business? The answer lies in the fact that input-output accounting systems follow standard firm-level accounting balance sheets: total assets and total liabilities are equal for a firm. In input-output analysis, profits are contained within the value-added entry and hence, the system represents a relatively complete picture of the transactions in an economic system. Note that for a transaction to be recorded, an exchange has to take place in the *marketplace*. Transactions that take place on the "black market" or in the "underground economy" are not recorded here; in developing economies these transactions could amount to a significant percentage of the total volume.

Links with National Accounts

Richardson (1972) has provided the link between this format and the national accounting framework articulated in Chapter 2. To accomplish this link, some variables need to be defined: Let X_{ij} be the flow of commodities from industry i to industry j on current account; f_{ik} is the flow of commodities from industry i to category k of final demand (this might be consumption or government for example); v_{mj} is the purchases from value added category m by industry j, and X_i is the total input (output) for sector i. The variables v and f may be disaggregated as follows:

$$v_j = P_j + W_j \qquad [3.1]$$

$$f_i = C_i + G_i + I_i + E_i \qquad [3.2]$$

where P and W represent profits and wages and salaries and C, G, I, and E are consumer, government, investment, and export sales. Thus, we may now write the row and column balances:

$$\sum_j X_{ij} + C_i + G_i + I_i + E_i = X_i \qquad [3.3]$$

$$\sum_i X_{ij} + P_j + W_j + M_j = X_j \qquad [3.4]$$

In this case, we have aggregated both types of imports and both types of exports into one category each, namely, M and E. If equations 3.3 and 3.4 are summed over all i and j sectors, we have:

$$\sum_i \sum_j X_{ij} + C_i + G_i + I_i + E_i = \sum_i X_i \qquad [3.5]$$

$$\sum_j \sum_i X_{ij} + P_j + W_j + M_j = \sum_j X_j \qquad [3.6]$$

It was noted earlier that the vectors of total input and output were equal; hence, the right-hand sides of equations 3.5 and 3.6 are equal. This provides for the expression:

$$\sum_i \sum_j X_{ij} + C_i + G_i + I_i + E_i = \sum_j \sum_i X_{ij} + P_j + W_j + M_j \qquad [3.7]$$

Again, note that the interindustry transactions, x_{ij} and x_{ij}, are contained on both sides of the equation; they are obviously equal and hence drop out. If we define C, G, I, E, P, W, and M as the vectors representing the variable elements in equation 3.7, we now have:

$$C + G + I + E = P + W + M \qquad [3.8]$$

Rearranging, we have:

$$C + G + I + E - M = P + W \qquad [3.9]$$

The left-hand side should look familiar: It is gross national product, whereas the right-hand side is gross national income. In the input-output framework, total final demand equals total value added plus imports. The latter two categories are known together as primary inputs. Thus, we have a strong link between national product and income accounts and the input-output model. Why are interindustry transactions ignored? Essentially, because their inclusion would amount to double-counting: The volume of flows or the number of intermediate transactions are not the major issue—rather, the amount of value created at each stage of production is of importance. This *is* included in the estimates for gross national product. Clearly, we can appreciate that at the regional level a comparable set of calculations will yield an estimate of gross regional product.

Regional Input-Output Model

Figure 3.2 represents the transactions table for a simple, hypothetical economy: To simplify analysis, each industry is shown to have an output of \$100 million. (There is no economic reason for this; in reality, the variation in levels of output will be substantial.) The comparison with Figure 3.1 reveals that interindustry transactions that occur *within the region* constitute \$298 million out of the total output (input) of \$500 million; however, the variations by industry are rather large. For example, industry 1 sells only \$33 million to other sectors, whereas sector 3 sells \$80 million. Similar variations may be seen in terms of purchases from other sectors.

Figure 3.2 represents an input-output table for a regional economy; our next task is to convert this to an analytical model. Let us aggregate the entries in columns 7 through 12 of final demand into one column (i.e., column 13); calling this entry f_i. Equation 3.3 may now be written:

$$\Sigma X'_{ij} + f_i = X_i \qquad [3.10]$$

The term X′ will be explained below. We will now make a number of important restrictive but necessary assumptions. First, assume that the demand for inputs is independent of the level of output. By this, we mean that the "recipe" for production (the percentage of total inputs required from each industry) does not vary with the scale of production. Second, the production system is such that no substitutions may be made. Hence, the proportionate use of inputs cannot be changed. These assumptions allow us to define a technical and a regional input coefficient:

$$a_{ij} = X_{ij}/X_j \qquad [3.11]$$

$$r_{ij} = X'_{ij}/X_j \qquad [3.12]$$

The coefficient a_{ij} represents the cents' worth of input purchased from sector i by sector j per unit of output of sector j; r_{ij}, on the other hand, provides an estimate of the proportion of the a_{ij} purchase that is made from firms located within the region. Hence, the distinction between the two coefficients lies in the distinction between purchases made irrespective of geographic origin (X_{ij}) and those made from within the region (X'_{ij}). Hence, we require a further assumption, namely that the propor-

Sectors	*1*	*2*	*3*	*4*	*5*
			Sectors		
1	0.21	–	0.09	0.03	–
2	0.01	0.08	0.07	0.29	–
3	0.03	0.20	–	0.50	0.07
4	0.31	0.02	0.38	–	0.03
5	0.10	0.25	0.26	0.01	0.04

Figure 3.3 Regional Input Coefficient Matrix

tion purchased within the region does not vary over different levels of production. For the analysis that follows, we will use expression 3.12. Rewriting this in terms of X'_{ij}, we have:

$$X'_{ij} = r_{ij} X_j \qquad [3.13]$$

Substituting this expression in equation 3.10 yields:

$$\sum_j r_{ij} X_j + f_i = X_i \qquad [3.14]$$

Figure 3.3 shows the values for the entries in the interindustry portion of the matrix in Figure 3.2, converted to regional input or regional requirements coefficients. These entries provide the proportion of the total requirements (namely the a_{ij}s) necessary to make $1 of output in the jth industry that come from industries within the region. As we have five industrial sectors, there will be five equations of the type shown in equation 3.14. In matrix terms, the system may be set up as follows:

$$\begin{bmatrix} r_{11} & r_{12} & r_{13} & r_{14} & r_{15} \\ r_{21} & r_{22} & r_{23} & r_{24} & r_{25} \\ r_{31} & r_{32} & r_{33} & r_{34} & r_{35} \\ r_{41} & r_{42} & r_{43} & r_{44} & r_{45} \\ r_{51} & r_{52} & r_{53} & r_{54} & r_{55} \end{bmatrix} \begin{bmatrix} X_1 \\ X_2 \\ X_3 \\ X_4 \\ X_5 \end{bmatrix} + \begin{bmatrix} f_1 \\ f_2 \\ f_3 \\ f_4 \\ f_5 \end{bmatrix} = \begin{bmatrix} X_1 \\ X_2 \\ X_3 \\ X_4 \\ X_5 \end{bmatrix} \qquad [3.15]$$

If we let R be the 5 × 5 matrix of r_{ij}s; X the 5 × 1 vector of total outputs (inputs) and f the 5 × 1 vector of final demands, equation 3.15 may be written in a more compact form that we may solve the simultaneous system in a more efficient manner.

$$RX + f = X \qquad [3.16]$$

If equation 3.16 is rearranged and factored, a solution for X may be derived:

$$f = X - RX$$

$$= (I - R)X$$

$$(I - R)^{-1}f = X \qquad [3.17]$$

where I is an identity matrix, a matrix with the value 1 along the main diagonal and zero elsewhere. If equation 3.15 is examined, it will be clear that the production of output in each industry theoretically involves the purchases of inputs from all other industries. In reality that will not be the case; an examination of Figure 3.3 reveals that sector 2 makes no purchases from sector 1 and sector 5 has no direct needs from sectors 1 and 2.[5] However, this should not be interpreted as implying that these sectors are not linked at all; the operative word here is the *direct* linkage. As we shall see later, firms that are not linked directly may be linked indirectly. For example, note that sector 2 makes no purchases from sector 1 but does purchase inputs from sector 3. This sector, 3, purchases inputs from sector 1. Hence, if the output of sector 2 expands, sector 1 will benefit in the second round of purchases. This may be shown diagrammatically in Figure 3.4. Note that the interactions become very complex and interwoven as the various rounds of spending and re-spending evolve. The analogy with the operation of the consumption effects in Figure 2.1 should be clear; after all, equation 3.17 can be rewritten:

$$(I + R + R^2 + R^3 + R^4 \ldots\ldots)f = X \qquad [3.18]$$

The various expressions of R represent the rounds of spending taking place in the economy; because R is a coefficient matrix, $R^t \to 0$ as $t \to \infty$, where t denotes the spending round. Hence, the contribution of each succeeding round diminishes. This part of the equation system in equa-

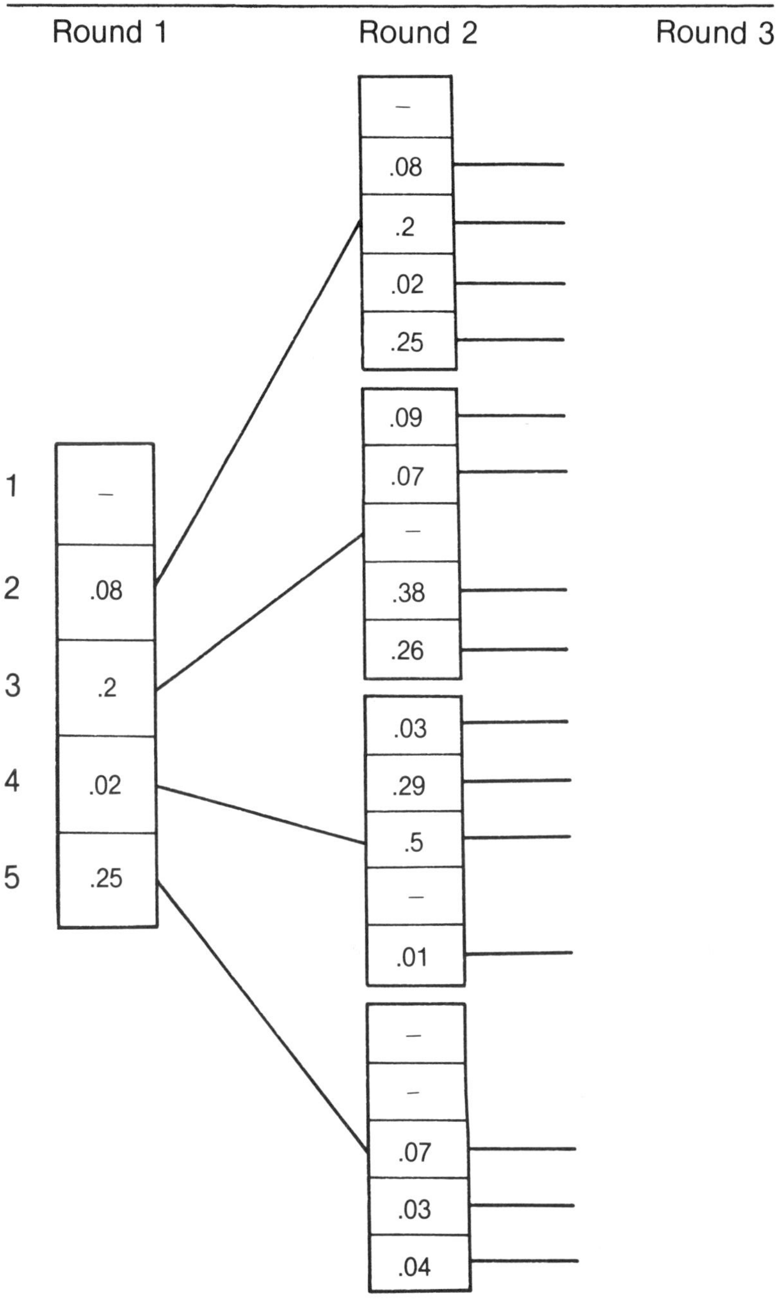

Figure 3.4 Rounds of Spending Imports for Sector 2

	Sectors				
	1	*2*	*3*	*4*	*5*
1	1.33	0.05	0.18	0.15	0.02
2	0.23	1.17	0.30	0.50	0.04
3	0.40	0.36	1.41	0.82	0.13
4	0.58	0.19	0.61	1.38	0.09
5	0.31	0.41	0.48	0.38	1.09
Multiplier	2.85	2.18	2.98	3.23	1.37

Figure 3.5 Leontief Inverse Matrix

tions 3.18 and 3.17 is thus the direct analogy with the multiplier articulated in Chapter 2. In input-output models, it is known as the Leontief Inverse Matrix or the Matrix Multiplier. The values for the regional example are shown in Figure 3.5. These entries provide the total requirements within the region for each industry to deliver $1 worth of output to final demand. Our suspicions that sectors might be linked indirectly is well illustrated here. As a result of all the transactions in the economy, note that sector 2 purchases 0.05 cent's worth of output from sector 1 in order to make a delivery of $1 to final demand.

What is the direct relationship of the entries in Figure 3.5 to the multipliers noted in Chapter 2? If the entries in a typical column of Figure 3.5 are summed, we have what is known as the output or column multiplier. Unlike the economic base model or the Keynesian system, we now have a multiplier *for each industry*. These are shown in the last row of Figure 3.5. Note that they vary rather substantially from one sector to another; hence, the use of an aggregate multiplier in an economic base model may provide a somewhat misleading impression of the likely impacts of any change in external demand on the regional economy. The reasons for these variations may be ascribed to (1) the degree to which industrial sectors are linked with each other (i.e., the number of nonzero entries in the matrix of interindustry transactions) and (2) the strength of those linkages (i.e., whether the linkages between sectors are of the same order of magnitude or dominated by one or two very large linkages). In Figure 3.3, one can see that sector 5 makes purchases from two other sectors (apart from itself) and these are not very large amounts. Accordingly, it has the lowest multiplier. Sector 4, on the other hand, is more intensely linked with the rest of the economy and has the largest multiplier.

At this stage, it would be erroneous to equate the size of the output multiplier with the importance of that sector in the regional economy. The multiplier tells us nothing about the *level* of sectoral output or its importance in terms of employment generation or income formation. We will deal with these issues in the next section.

Income and Employment Multipliers

Not only do industries make purchases from other sectors, they also make purchases from the labor force. The next task is to calculate the various income multipliers associated with these purchases; there are three and possibly more types of income multipliers that can be obtained. In this section, we will restrict ourselves to the two most commonly identified, namely, the Type 1 (direct and indirect) and the Type 2 (direct, indirect, and induced). To do this, we will first derive an additional row of coefficients, these being the entries shown in row 7 of Figure 3.2. Assume for the moment that all the entries in the value added row are for wages and salaries. If these entries are divided by the relevant sectoral output, we can obtain a vector showing cent's worth of labor input per unit of output. This is shown below:

Sector Number	1	2	3	4	5
Input of Labor	0.2	0.4	0.1	0.17	0.4

Here we see rather substantial differences in the purchases of labor by sector. These entries will now be represented by a row vector V. If this vector is multiplied against the Leontief Inverse matrix, we will obtain a matrix of direct and indirect income changes:

$$V(I - R)^{-1} \qquad [3.19]$$

If these entries are then divided by the direct income changes, V, we have what is known as the Type 1 Income Multiplier (M1):

$$M1 = V(I - R)^{-1}\hat{V} - 1 \qquad [3.20]$$

where the $\hat{}$ indicates a vector expressed as a diagonal matrix. The procedure may be demonstrated with reference to sector 1; each entry in the first column of Figure 3.5 will be multiplied by the corresponding row entry detailing the labor input per unit of output (shown as the first row of Figure 3.6).

	Sectors				
	1	*2*	*3*	*4*	*5*
(1)	0.20	0.40	0.10	0.17	0.40
(2)	0.62	0.72	0.59	0.69	0.48
(3)	0.42	0.31	0.49	0.52	0.08
(4)	3.1	1.78	5.9	4.1	1.20

NOTE: Row (1) value-added entry of Figure 3.2 shown as a coefficient (V/X); (2) direct and indirect income change; (3) indirect income change (2) – (1); (4) type I multiplier, (2) ÷ (1).

Figure 3.6 Leontief Inverse Matrix

$$
\begin{aligned}
1.33 \times 0.20 &= 0.26 \\
0.23 \times 0.40 &= 0.09 \\
0.40 \times 0.10 &= 0.04 \\
0.58 \times 0.17 &= 0.09 \\
0.31 \times 0.40 &= 0.12 \\
\text{Total} &= 0.62
\end{aligned}
$$

The summation above may not equal 0.62 exactly because of rounding error; the results of this manipulation for all sectors are shown in Figure 3.6. Again, caution should be exercised in ascribing importance to size in terms of multiplier values. A sector with a large entry in V and with a high level of output may offset the fact that its income multiplier may be relatively small. Because industries 3 and 4 are so highly linked in the system, their indirect income effects are very large in comparison to the direct income effects—hence, the income multipliers are very large.

The income picture is not really complete because we have not taken into account the fact that wages and salaries received by local employees may be spent on local goods and services, thereby generating additional output and, hence, additional income. There are a number of ways in which these effects can be calculated. The most direct way is to expand or augment the direct coefficients' matrix (Figure 3.3) to include the additional row V that we have already defined and an additional, complementary column. This column is the vector of consumption coefficients by sector; in a sense, it represents a disaggregation of the average propensity to consume by all households in the region (i.e., column 7 of Figure 3.2). This consumption is restricted to the output of goods and services produced within the region; imports are shown in

	Sectors					HH
	1	*2*	*3*	*4*	*5*	*6*
1	0.21	–	0.09	0.03	–	0.18
2	0.01	0.08	0.07	0.29	–	0.15
3	0.03	0.20	–	0.50	0.07	0.03
4	0.31	0.02	0.38	–	0.03	0.07
5	0.10	0.25	0.26	0.01	0.04	0.05
HH 6	0.20	0.40	0.10	0.17	0.40	0.01

Figure 3.7 Expanded R Matrix

rows 8 and 9 of column 7 of Figure 3.2. The augmented matrix (R′) is shown in Figure 3.7; the solution is now obtained in a similar way, except that f′ does not contain the consumption account:

$$(I - R')^{-1}f' = X \qquad [3.21]$$

The multiplier matrix, $(I - R')^{-1}$, is shown in Figure 3.8. Type 2 income multipliers, M2, are simply the division of the household row entries of the augmented Leontief Inverse of Figure 3.8 by the direct income entries, V. These are shown in Figure 3.9 together with a comparison with the M1 multipliers. Several authors have noted, and subsequently proved, that, for any matrix, there is a constant relationship between the M1 and M2 multipliers. As this constant can be derived without the inversion of the augmented matrix, once the M1 multipliers are known, the M2 values can be calculated rather easily. (For a proof, see Miller and Blair 1985, p. 143.)

A final note on income multipliers should be made; income *generated* within a region may not be equal to income *retained* and then spent in the region. Labor force commuting across regional boundaries, repatriation of income to other regions and various taxation levies applied by state and national governments may all serve to reduce the size of the income pool expended in the region. On the other hand, additional non-wage and salary income (dividends, interest payments, and government transfers) may enhance the pool of income earned from employment. These accounting issues have been addressed in many recent more sophisticated regional social accounting models (see Batey and Madden 1983, for an example).

Finally, if we assume that the levels of employment in an industry are closely related to output, such that an employment/output ratio can be

Sectors	*Sectors* 1	2	3	4	5	*HH* 6
1	1.59	0.34	0.43	0.43	0.21	0.41
2	0.48	1.46	0.54	0.78	0.23	0.41
3	0.62	0.61	1.62	1.07	0.30	0.36
4	0.82	0.47	0.84	1.65	0.28	0.39
5	0.52	0.65	0.68	0.62	1.25	0.34
HH 6	0.3	1.07	0.89	1.05	0.72	1.50

Figure 3.8 Leontief Inverse Matrix with Households Endogenous

defined for all levels of output, then the entries in the input-output system can be converted to employment terms to yield employment multipliers. The procedure is as follows; rewrite equation 3.10 in matrix form:

$$X'i + f = X \qquad [3.22]$$

where X′ is the matrix of interindustry flows within the region and i is an identity vector that we use to sum across the rows of matrix X′. If e is defined as the employment vector showing employment by sector, then the expression eX^{-1} provides the employment output ratios (number of jobs per $million of output). If all the entries in equation 3.22 are multiplied by this expression, we have:

$$\hat{e}\hat{X}^{-1}X'i + \hat{e}\hat{X}^{-1}f = \hat{e}\hat{X}^{-1}X \qquad [3.23]$$

The right-hand side becomes e, because $X^{-1}X$ is equal to the identity matrix; that fact can also be used to replace i on the left-hand side with $e^{-1}e$. Hence, we now have:

$$\hat{e}\hat{X}^{-1}X'\hat{e}^{-1}e + e\hat{X}^{-1}f = e \qquad [3.24]$$

Factoring and simplifying, we have:

$$\hat{e}\hat{X}^{-1}f = e - \hat{e}\hat{X}^{-1}X'\hat{e}^{-1}e$$

$$= (I - \hat{e}\hat{X}^{-1}X'\hat{e}^{-1})e$$

$$(I - \hat{e}\hat{X}^{-1}X'\hat{e}^{-1})^{-1}\,\hat{e}\hat{X}^{-1}f = e \qquad [3.25]$$

	Sectors				
	1	*2*	*3*	*4*	*5*
(1)	0.20	0.40	0.10	0.17	0.40
(2)	0.93	1.07	0.89	1.05	0.72
(3)	4.67	2.68	8.99	6.19	1.89
(4)	3.1	1.78	5.9	4.1	1.20

NOTE: Row (1) value-added coefficient derived from Figure 3.2; (2) household row of Figure 3.8—direct, indirect, and induced income change; (3) type II multiplier, column (2) − (1); (4) type I multiplier, from Figure 3.6.

Figure 3.9 Derivation of Type II Multipliers and Comparison with Type I

The inverse matrix in equation 3.25 is the direct analogy to the Leontief inverse matrix in dollar terms; all the entries are now expressed in employment terms. Given a change in final demand, we can determine the level of employment required, directly and indirectly, in each industry. The expression $\hat{e}\hat{X}^{-1}X'\hat{e}^{-1}$ converts the cents per dollar coefficient matrix shown in Figure 3.3 into one showing employment coefficients, e_{ij}s, the employment required from industry i per employee in industry j to support output in industry j.

Summary

Now that the conversion of the input-output table to a model has been demonstrated, the model can be used for analytical purposes. The reader should be wary of inferring too much from the rankings of employment and income multipliers. For example, many labor intensive sectors have low employment multipliers simply because the denominator of the multiplier (the direct effects) is relatively large. In absolute terms, though, these sectors may generate the largest volume of employment. In the next chapter, some simple examples will be provided prior to some discussion about the ways in which an input-output model might be constructed from other than survey data. Chapters 6 and 7 will extend this simple, single-region framework spatially (into the interregional context) and sectorally (by specifying different income groups, different types of labor and so forth).

4. APPLICATIONS

Now that an input-output model has been developed from the survey data and the input-output table, we can use the system to undertake some relatively straightforward applications of the model. In this chapter, two illustrations will be provided; the first examines the impact on employment and output of a change in a federal government's programs. For example, we might wish to consider the effects of a shift away from defense to social programs. The second example shows how the input-output model might be used to undertake a cost-benefit analysis of a project. Prior to the discussion of these in-depth examples, some comments will be made on the range of regional input-output applications.

The early applications of regional input-output analysis focused almost exclusively on impact analyses, for example, the effects of government programs on a regional economy. More creative uses were found—for example, the impact of sports franchises on metropolitan economies, the employment and income generated by large institutions such as universities, and the impacts of new transportation facilities on regional economies. Increasing interest in resource scarcity, and environmental and energy problems fostered a whole new series of applications for input-output models. In many cases, the regional models were linked with other analytical systems or recast to provide more flexible forms of analysis. Within these categories, one finds applications of air pollution abatement programs, the effects of water shortages on regional economic growth, and development and many applications exploring the effects of disruptions in energy supply on various regional economic indicators.

Regional input-output models have been used for policy simulation, for forecasting employment, output and income, and as components in integrated modeling ventures. A great deal more detail is provided in Miller and Blair (1985), especially on the applications of input-output analysis in the solution of energy and environmental problems.

Impact Analysis with an Input-Output Model

Let us assume that the federal government is considering a cut of $10 million in defense spending in the regional economy and reallocating the funds either to a nondefense-oriented set of programs (e.g., social welfare, education) or to consumers in the form of a reduction in taxes. Hence, in the first case, government expenditure in the region will remain the same ($10 million). Should one therefore conclude that the net impact on the region's economy will be zero? If we had been working with a simple economic base model, that would have been a correct assertion. However, even though the total expenditure by the federal government in the region will remain the same, the impacts will not

	(1) HH	(2) Δ HH	(3) Defense	(4) Δ D	(5) Nondefense	(6) Δ ND
1	30	0	1	– 4	–	–
2	25	0	0	– 2	–	–
3	7	+ 2	2	– 2	6	+ 2
4	15	+ 3	–	–	13	+ 2
5	10	+ 1	–	–	16	+ 3
HH 6	3	+ 1	2	– 2	12	+ 3
Imports 7	80	+ 3	–	–	–	–
Total	170	+10	5	–10	47	+10

NOTE: Imports rows of Figure 3.2 have been combined. HH, households; D, defense; ND, nondefense.

Figure 4.1 Reallocation of Final Demand Consequent Upon a Reduction in Defense

necessarily be identical. The reasons lie in the differences in the *allocation* by sector of the two budget reallocations in comparison with the original defense-related expenditure. It is unlikely that the same goods and services will be needed to support a defense program as will be needed for a social welfare program or the consumption needs of local consumers. For this reason, we should not expect similar impacts from the reallocation from defense to consumer spending even though, once again, the total amount of final demand being allocated in the region is the same.

Figure 4.1 shows the reallocation of final demand from defense to nondefense and from defense to households. In the allocation to households, note that the $10 million was not distributed proportionately to the original vector of expenditures. Empirical evidence suggests that consumers allocate *additions* to income differently than their *average* income. Thus, although the original vector might be regarded as an expression of average propensities to consume, the distribution of the $10 million is assumed to follow the dictates of a vector of marginal propensities to consume. Similarly, the vector of additions to nondefense spending reflects a slightly different reallocation from the existing distribution, although one that is not as significantly different in appearance as those between the average and marginal propensities to consume by households.

Figures 4.2 and 4.3 show the results of the impact analyses; using the employment/output ratios shown in Figure 4.2, the employment impacts were also calculated as well as the output effects. As we sus-

Sectors	Decrease Output	Defense Employment	Increase Output	Nondefense Employment	Increase Output	Household Employment	Employment per $1 Million Output
1	− 5.78	−1056	+ 0.72	+ 154	+ 0.83	+ 166	200
2	− 3.86	−1930	+ 1.72	+ 860	+ 2.14	+1070	500
3	− 5.14	−1542	+ 4.85	+1455	+ 5.41	+1623	300
4	− 3.90	−1950	+ 4.25	+2125	+ 5.45	+2725	500
5	− 3.02	−1208	+ 4.99	+1996	+ 3.19	+1276	400
Total ($)	−21.70	−7686	+16.53	+6590	+17.02	+6860	

Figure 4.2 Employment and Output Effects of Changes in Final Demand

pected earlier, even though the total amount of money being spent by the various sectors of final demand was unchanged, the reallocations severely altered the outcomes on a sector-by-sector basis. The gross effects are shown in Figure 4.2 and the net effects are summarized in Figure 4.3. Sectors 4 and 5 were net gainers from a reallocation to either households or nondefense spending, whereas sector 3 gained from the household but not the nondefense reallocation. Sectors 1 and 2 would suffer losses from both reallocations. Because the employment data are linearly dependent on the output figures, these results hold for both employment and output. The magnitude of the changes varied by sector rather appreciably. Notice that sector 5 gained far more from a nondefense reallocation than from a consumer spending addition; in part, this is due to the fact that $3 million of the reallocation would be "lost" to the regional economy if added to consumers' income by reason of import purchases. The overall effects led to decreases in available jobs in the region and a loss of production. Several comments need to be made here; first, no consideration is given to the possibilities that some sectors may not be able to adjust their production from defense to nondefense or consumer goods instantaneously. Second, the employment data reflect *average* employment/output ratios; some sectors may be able to produce less output by curtailing employment opportunites in a greater than proportionate manner, thus creating additional layoffs. Third, nothing is revealed in this analysis about the various skill categories required; for some occupations, the demand for their skills may actually arise during a reallocation, whereas for others the demand may be drastically reduced. Hence, the overall employment impacts may hide significant dislocations and disequilibria in supply-demand relatonships.

A final word of caution: one should not infer from this analysis that defense spending is essential to the maintenance of regional economic vitality. This example uses data that may or may not reflect reality.

	Defense to Nondefense		Defense to Household	
Sectors	Output	Employment	Output	Employment
1	−5.06	− 902	−4.95	−890
2	−2.14	−1070	−1.72	−860
3	−0.29	− 87	+0.27	+ 81
4	+0.35	+ 175	+1.55	+775
5	+1.97	+ 788	+0.17	+ 68
Total ($)	−5.17	−1096	−4.68	−826

Figure 4.3 Net Gains (Losses) from Reallocation of Final Demand

Furthermore, nothing is revealed about the possible longer-run implications of such changes. Some local firms, faced with declining demand, may make some changes in their production process or product lines in the hopes of securing new markets within the region and elsewhere. The input-output model cannot hope to answer all the questions related to impacts of the kind demonstrated here. One major effect not considered, of course, is the role of migration. In regional economies, downturns and upswings in the business cycle can have pronounced effects on the volume of in and out migration. The expectations on the part of individuals, with respect to the duration of unemployment, will also play a critical role in the "mover/stayer" equation. However, the sectoral detail afforded by the input-output model does provide an important source of information not available from other, more aggregated methods of regional analysis. Similar studies can be undertaken, for example, to measure the impact of a new firm (even a sports franchise) on a regional economy or the effect of the closure of an existing plant. In the latter case, the mobility of the newly unemployed workers will play a crucial role in maintaining the remaining levels of activity in the region. This issue will be addressed in Chapter 7.

Using an Input-Output Model in Project Appraisal

In many economies, one of the more important objectives associated with economic development is the need to narrow the differences in welfare between regions. In a number of cases, the index used for comparison is the level of income per capita. Achieving this goal requires careful selection of policy instruments designed to steer the allocation of resources in such a way that the less prosperous regions achieve some possibility for improving their welfare at a faster rate than

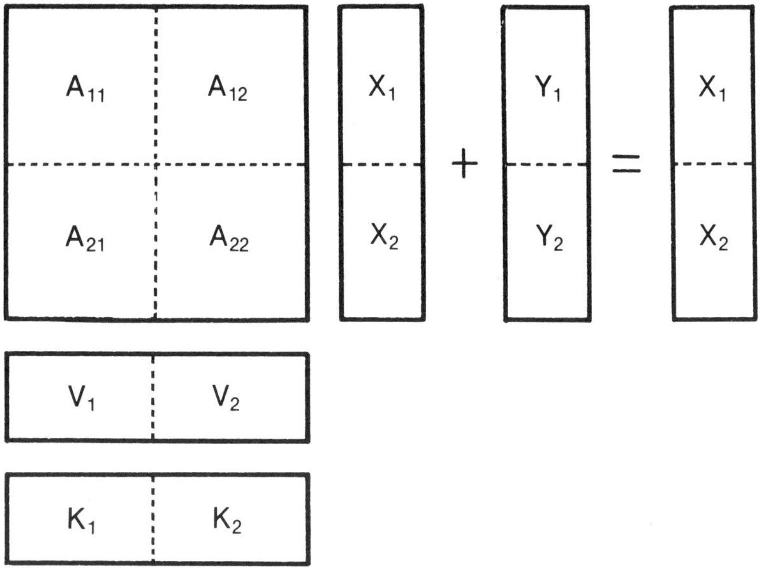

Figure 4.4 Tinbergen Semi-Input-Output Formulation

the more prosperous regions. The choice of these policy options would form the subject of another monograph and will not be considered here; let us assume that the country in question has decided that the availability of large supplies of coal in the region might provide an opportunity for the penetration of markets in other regions and other countries, especially in the face of rising petroleum prices. However, the project will require some initially large capital investment in machinery, transportation systems, and so forth. Can the project be justified? If we restrict ourselves to economic issues, cost-benefit analysis can be used to help answer the question about the wisdom of this project vis-à-vis several others. The linkage of input-output models with cost-benefit appraisal techniques was first suggested by Tinbergen (1966) and has subsequently been utilized in a number of different contexts (see Kuyvenhoven 1978; Karunaratne 1976; and for a regional application, Bell, Hazell, and Slade 1982).

Figure 4.4 shows the general form of the modification that is necessary to the input-output table to achieve this linkage. Sectors are classified into (1) tradeables, those sectors that engage in markets outside the

						Total Initial Demand Change*
Mining	.21	.0	.09	.03	.0	15
Iron & Steel	.01	.08	.07	.29	.0	0
Elec. Eng.	.03	.20	.0	.50	.07	0
Bus. Serv.	.31	.02	.38	.0	.03	0
Trans.	.10	.25	.26	.01	.04	0
V	.2	.4	.1	.17	.4	
K	.13	.15	.10	.05	.10	

$$[I - R_{NN}]^{-1} = \begin{bmatrix} 1.27 & 0.51 & 0.46 & 0.47 \\ 0.29 & 1.39 & 0.60 & 0.45 \\ 0.51 & 0.84 & 1.43 & 0.37 \\ 0.04 & 0.12 & 0.09 & 1.08 \end{bmatrix}$$

$$K_{NN}\,[I - R_{NN}]^{-1} R_{NJ} = 0.0752$$

$$V_N\,[I - R_{NN}]^{-1} R_{NJ} = 0.256$$

*Associated with project.

Figure 4.5 Empirical Implementation of the Tinbergen Model

region for the sales of their goods and (2) nontradeables, those sectors whose markets lie within the region. The subscripts T and N indicate these distinctions; the regional coefficient matrix, R, of Figure 3.3 is now partitioned into four submatrices. A twofold partition, into tradeables and nontradeables, for final demand and output, is also made. An additional vector is included in the system, the vector of capital coefficients, reflecting the needs of capital account per unit of input.[6] Figure 4.5 shows the rearrangement of the input-output table to accommodate the distinctions introduced by Tinbergen. In undertaking the project appraisal, we will assume that the final demand for nontradeables is zero; this is not unreasonable, as we wish only to record the effects of the project, not the totality of the economic activity in the region. Hence, $f_n = 0$. Solving the two equations from Figure 4.4, we have:

$$R_{TT}X_T + R_{TN}X_N + f_T = X_T \quad [4.1]$$

$$R_{NT}X_T + R_{NN}X_N + f_N = X_N \quad [4.2]$$

where R_{TT} and R_{NN} represent the interindustry coefficients among the tradeable and nontradeable sectors, respectively; R_{TN} and R_{NT} show the transactions between tradeables and nontradeables. We will further assume that, as far as this project is concerned, there will be no sales from tradeables to nontradeables (because the coal will be exported). Hence, matrix R_{TN} will be zero and thus equation 4.1 reduces to:

$$(I - R_{TT})^{-1} f_T = X_T \qquad [4.3]$$

Bearing in mind that there are no sales to nontradeable final demand, equation 4.2 becomes:

$$(I - R_{NN})^{-1} R_{NT} X_T = X_N \qquad [4.4]$$

Let us identify the sector in which the coal mining takes place as sector j, then the cost-benefit criterion for this project will be:

$$T_J = \frac{K_J + K_N (I - R_{NN})^{-1} R_{NJ}}{V_J + V_N (I - R_{NN})^{-1} R_{NJ}} \qquad [4.5]$$

Because each of the terms in equation 4.5 is multiplied by X_T, the term drops out. The numerator represents the direct capital cost of the project (K_J) and the direct and indirect capital demands that would be placed on the nontradeable sectors as a result of this project. These demands might include additional highways and railroads, telephone systems, new infrastructures to house additional workers and their families. The denominator provides the benefits; these are divided into the direct value added created by the project (V_J) and the direct and indirect income consequent upon the expansion of demand to support the project through the sales of nontradeable goods to the tradeable sector. One would hope that the ratio would be less than one, indicating that the project benefits exceeded the costs. In the example shown in Figure 4.5, we assume that the project will incur a capital cost of approximately \$2 million for an annual output of \$15 million; we will assume that the value added ratio in the new project will be the same as in the mining sector as a whole—namely \$0.20 per dollar of output. Thus, $K_J = 2/15$ or 0.13 and $V_J = 0.20$. Manipulation of equation 4.5 yields:

$$T_J = \frac{0.13 + 0.0752}{0.20 + 0.256} = \frac{0.2052}{0.4560} = 0.45 \qquad [4.6]$$

In this case, the benefits exceed the costs rather substantially.

However, this form of appraisal is not a complete picture of the possible contributions of the project to the regional economy. Consider that it only takes into account costs and benefits in monetary terms. Nothing is stated about the employment creation benefits. Typically, one might expect multiple criteria to be applied to project evaluation, rather than the reliance on one measure alone (see Nijkamp and Spronk 1981).

Second, the denominator of equation 4.5 includes the component, $V(I - R_{NN})^{-1}$, which should look familiar. It is, of course, the direct and indirect income created in the nontradeable sectors (see equation 3.19). However, as we noted in Chapter 3, income creation is only one part of the cycle of activities; consideration also has to be given to the expenditure effects of this income. Added to this income for purposes of expenditure effects should be V_J, the income generated directly by the project. In all likelihood, these effects may be much greater than the interindustry linkage effects. In addition, no attempt has been made to consider the benefits and costs over a time horizon; consideration should be given to the *discounted* stream of net benefits that would provide some stronger indication of their opportunity costs. The formulation above may be unrealistic as far as capital costs are concerned because the initial capital costs and the ongoing capital costs may bear a very different relationship with total output. Similarly, many large-scale development projects may create significant employment opportunities during the construction phase of a project only to yield a very small number of jobs of a permanent nature once the project is up and running. This has been especially true in many resource development projects in the western United States. The "boom town" phenomenon associated with oil shale development is one example of massive short-run construction employment peaks followed by substantially lower, longer-run employment impacts. The necessity for inclusion of consideration for the time-phasing of projects in the input-output modeling system would seem to be a high priority modification.

One other major consideration has been ignored in this analysis; the existence of interregional effects may serve to bias the cost-benefit ratio. For example, if there are significant impacts outside the region, the level of income may rise faster there than in the region itself, in conflict with the avowed aim of the project to reduce income differentials between the regions. The cost appraisal technique would need to be integrated more fully with the nation's goals and objectives so that a decision to undertake a project would not be taken purely on the grounds of the value of T_J. As the internal complexity of the regional economy increases, the distinctions between tradeables and nontradeables may become less apparent. In this case, the alternative proposed by Herman et al. (1969) and Batten and Andersson (1983) may prove more attractive. They classified industries hierarchically (international, interregional, and regional) and were able to examine the interregional implications of changes in national policies such as the role of import substitution initiatives. Bell et al. (1982) categorized support or infrastructure activities as a third category to complement the distinctions between tradeables and nontradeables.

Summary

These two examples provide some measure of the scope for the use of the input-output model. Impact analysis and project appraisal have played important roles in decision making with reference to the allocation of federal and state resources. The input-output model provides some guidance as to the possible outcomes for proposed programs and, thus, offers policymakers the option of considering several alternatives prior to committing resources to any one program. Input-output models are, unfortunately, very expensive to construct from survey-based data. In the next chapter, some alternative, nonsurvey methods will be evaluated.

5. CONSTRUCTION OF REGIONAL INPUT-OUTPUT MODELS

The costs of constructing a survey-based regional input-output model may well involve expenditures in excess of one or two million dollars. Although national-level models have become an accepted part of the data collection and assembly processes associated with national income and product accounts, the same has not been true at the regional level. As a result, the development of input-output tables at the subnational level has relied on two alternative options, nonsurvey and partial survey techniques. These options will form the focus for the discussion in this chapter.

Nonsurvey Techniques

The existence of sets of input-output tables at the national level has provided for the development of non-survey-based regional input-output tables. In a very small number of cases, some regional tables have been constructed by modifying tables assembled from survey data for other regions. However, the major efforts have been directed toward the modification of national tables.

In the most common situations, the existence of the national tables has been complemented by very limited information at the regional level. These data usually constitute the various censuses of manufacturing, wholesale and retail trade, transportation, and agriculture supplemented by data available in *County Business Patterns* and aggregate time-series data at the county and state levels assembled by state and federal agencies such as the Bureau of Economic Analysis of the U.S. Department of Commerce. From these data, we are usually able to construct a vector of total output for a set of industries in the region; the general consensus of opinion is that greater sectoral disaggregation is preferable. The next steps involve the use of some modifying techniques to transform the national table to a regional one. These techniques involve the estimation of various quotients or ratios, many of which have been applied in other contexts in the regional analysis literature. In this chapter, we will review just a small selection of them; more detail may be found in Smith and Morrison (1974).

In earlier chapters, we noted the distinction between the set of technical coefficients, the a_{ij}s, and the regional requirements coefficients, the r_{ij}s. In operating with the nonsurvey techniques, an assumption is usually made to the effect that:

$$a_{ij}(r) = a_{ij}(n) \qquad [5.1]$$

namely, that the technology used in comparable industrial sectors is the same at the national (n) and regional (r) levels. This assumption thus precludes possible differences in the age of capital stock, the size mixture of firms within a sector, differences in technology, and possible variations in product mix. However, in the absence of any information about many of these characteristics at the regional level, one is left with very few options but to adopt a very conservative strategy, namely, one in which a minimum of speculation is applied to the modification process. As a result, the strong assumption needs to be adopted that the distributions of many of these characteristics concerning firms are similar at the national and regional levels. The various nonsurvey techniques seek to modify the national technical coefficients to produce a set of regional requirements coefficients:

$$r_{ij} = a_{ij}(n)\ q_j \qquad [5.2]$$

where q_j is a quotient or ratio of some kind. This quotient may be applied uniformly across all entries in a row or down a column or it may be developed as q_{ij} and applied separately to each individual coefficient. The choice of technique must be rooted in theory; in the development of the economic base model, similar problems arose in the estimation of the proportion of activity that was basic as opposed to nonbasic when survey data were not available. In this case, a popular nonsurvey technique that has been used is the *location quotient*.

Equation 5.3 shows the formula for the estimation of the location quotient for industry i in region r:

$$lq_i = [X_i(r)/\sum_i X_i(r)] / [X_i(n)/\sum_i X_i(n)] \qquad [5.3]$$

In essence, the proportionate share of industry i in the region is compared with the share of i in the nation. The data used for the Xs usually constitute either employment or output figures. Once the quotient has been obtained, it is applied in the following fashion:

$$r_{ij} = \begin{cases} a_{ij}(n) & \text{if } lq_i \geqslant 1 \\ a_{ij}(n)\, lq_i & \text{if } lq_i < 1 \end{cases} \qquad [5.4]$$

The reasons for this application may be stated in terms of the expectation that any industry, i, will be able to supply the demands placed upon it by all other industries in the region. If the industry's representation in the region is greater than that observed in the national data (i.e., the location quotient is greater than one), then there is a high probability that the industry will be able to meet all local demands. However, if the reverse is true, then in all likelihood, only a portion of the local demands will be able to be supplied from the local industry (the remaining demands will have to be imported). The share of the total demands supplied locally is thus reduced in accordance with the size of the location quotient. If $lq_i = 0.83$, then only 83 percent of the local needs of the output of industry i will be supplied within the region. If the industry is not present in the region, then all coefficients in that row would be zero. Note that no matter how much larger than one the location quotient might be, the national technical coefficient serves as an upper bound. The larger, more diversified the region, the greater the expectation that the region will be able to supply most of its own needs. This follows from the fact that as the size of the region approaches the nation, the values of the location quotients will approach one (from above and below)[7] and, hence, the regional coefficients will converge toward the national coefficients. There are several techniques of this kind, relying on very aggregate data for their estimation.

Stevens et al. (1983) have developed a technique that attempts to use regional theory to develop an appropriate way of modifying national technical coefficients. Following some earlier work by Stevens and Trainor (1976) that showed that the regional purchase coefficients were an important source of error in regional models, they developed a technique that may be summarized as:

$$R = \hat{P}\, A(n) \qquad [5.5]$$

where $\hat{P}$ represents a diagonalized matrix of regional purchase coefficients for industry i. Instead of using location or other quotients to estimate P, an estimating equation was developed in which P was regressed on various quotients such as the wages in the region compared to the nation, output in the region compared to the nation, and so forth. A comparison of this technique with known survey-based coefficients yielded acceptable results although the differences varied considerably by industry.

The question that now arises is the degree to which we are able to produce nonsurvey tables that are within acceptable levels of accuracy. One of the major unresolved debates in regional input-output analysis focuses on this very subject. One of the reasons for the discord centers on the measurement of accuracy; should one expect a table to be replicated accurately on a coefficient-by-coefficient basis, or on the basis of the use of the table in developing the input-output model and the associated multipliers? Jensen (1980) has tried to adjudicate this debate by suggesting that many of the problems revolved around different perspectives on accuracy. For example, many applications of the input-output table focus on the use of the multipliers; their accuracy is the major concern and thus we may refer to this as an example of a focus on holistic accuracy. On the other hand, a specific industry impact analysis may require very careful estimation of the outcomes on a sector-by-sector basis; this concern reflects a focus on partitive accuracy. The issue, though unresolved, presents an important point of contention in the continued development of nonsurvey techniques, for although there is general agreement that they are not very satisfactory, there is also a clear appreciation of the fact that funds for the development of full survey-based models are unlikely to be forthcoming in the future. A possible compromise, partial survey techniques, provides some expectation for achieving acceptable levels of accuracy under conditions of limited regional information and with minimal funding.

Partial Survey Techniques

In this context, one is faced with the possibilities for developing an input-output table with a funding level that would allow the application of techniques that are a little more sophisticated than the ones described above. What alternatives are available? Three alternatives will be presented for consideration. All require some survey data collection but not on the scale envisaged for the complete table development described in Chapter 2.

THE RAS OR BIPROPORTIONAL TECHNIQUE

The RAS or Biproportional technique was developed by Stone (1963) to provide a method for updating national input-output tables over short time periods. Regional researchers have used the technique, and variations on it, to adjust national coefficients to the regional level. The general expression is:

$$R = f[A(n), u(r), v(r), X(r)] \qquad [5.6]$$

where A(n) is the national technical coefficient matrix and u(r), v(r), and X(r) are regional vectors of total intermediate outputs, intermediate inputs, and total outputs, respectively. Because we know that A(n) is available and that X(r) can be developed from various regional data, the remaining vectors for which we need regional data are u and v. Attempts to estimate these vectors using nonsurvey techniques have proved very unreliable. Hence, a sample of firms would be required to be surveyed and asked two questions: the volume of sales made to other firms within the region and the volume of purchases made from other firms within the region. These data would then form the basis for the estimation of u and v. Stone showed that, given these data, the following was true:

$$R = \hat{r}\, A(n)\, \hat{s} \qquad [5.7]$$

where ^ indicates a diagonal matrix. Both Stone and Leontief have suggested interpretations for the values of r and s (see Bacharach 1970).

Using Stone's interpretation (suitably translated to the regional context), we may consider the elements of r, r_i, to measure the extent to which the industry i at the regional level is unable to supply the inputs required by all other industries to meet their requirements. In this sense, it operates in a similar fashion to the location quotient. It is also independent of any given entry in row i as it modifies all elements of the ith row in a proportional fashion. The elements of s, s_i, are a measure of the fabrication effect in the production of j—the extent to which the industry that produces j at the regional level uses more or less intermediate inputs per unit of gross output than the corresponding industry at the national level. Note two differences in the application of this technique compared to the location quotient approach. First, there is no restriction of the kind proposed in equation 5.1, namely the upper bound on input coefficients implied by national technology. Second, each coefficient is modified, potentially, in a different manner because we may see that equation 5.7 operates on an element-by-element basis:

$$r_{ij} = r_i a_{ij}(n) s_j \qquad [5.7']$$

Although the r_is and s_js operate uniformly on a row (column) basis, the interactive effect is likely to be different for each coefficient. Several outcomes are possible; the net effect might be neutral (i.e., $r_{ij} = a_{ij}$) or it could yield regional coefficients that are larger or smaller than the national values.

Two other aspects of this technique should be articulated. Unlike the location quotient procedure, the RAS technique guarantees that the resulting coefficient matrix will satisfy the constraints, u(r) and v(r). In addition, it will also yield a matrix that is as close as possible to the prior matrix. This makes sense when the technique is used to update a matrix from time t to t + n; does it also make sense at the regional level? In the absence of specific coefficient information, the adoption of a procedure that ensures that the margins are estimated accurately, and provides for essentially the minumum disturbance of the national coefficients to achieve this, may be justified. The operation of the technique will illuminate these two points rather well. Let numerical subscripts refer to successive approximations of the vector or matrix associated with them. Given the data available in equation 5.6, an estimated vector of intermediate outputs is obtained using the national coefficients and known regional outputs:

$$u_1 = A(n)\, X(r) \qquad [5.8]$$

These estimates are adjusted to conform to observed values of u(r) via adjustments to the matrix A(n). A new matrix, A_1, produced:

$$A_1 = \hat{u}(r)\hat{u}_1^{-1}A(n) \qquad [5.9]$$

This new matrix is now used to obtain estimates of intermediate input, v_1, and matrix A_1 is further adjusted to a new matrix A_2 to ensure equality with observed intermediate inputs, v(r).

$$v_1 = \hat{X}(r)A_1'i \qquad [5.10]$$

$$A_2 = A_1\hat{v}(r)\hat{v}_1^{-1} \qquad [5.11]$$

The process returns to equation 5.8 where u_2 is now estimated using:

$$u_2 = A_2X(r) \qquad [5.12]$$

and so on through equation 5.11. Equations 5.8 through 5.11 represent one complete iteration. Empirical evidence suggests that the process converges rapidly, usually within ten iterations. Notice that each phase involves adjustment of the coefficient matrix so that when it is transformed into the transaction matrix, it conforms to the known vectors of either intermediate input or output. Thus, accuracy with reference to these vectors is assured. Because the adjustment process operates on the A matrices, the adjustment process is conservative, making only the minimally necessary adjustments to ensure agreement with the vectors u(r) and v(r). Miernyk (1976), among others, has been critical of the RAS technique applied to the development of regional input-output models from national models. Hewings (1977) has shown that the choice of the prior matrix is critical; a random-number generated matrix was adjusted via the RAS technique to produce a holistically acceptable estimate of the survey-based input-output matrix for the state of Washington. However, inspection of the individual coefficients revealed little correspondence between the observed and estimated values.

MOST IMPORTANT COEFFICIENT ESTIMATION

Several researchers (notably Jensen and West, 1980) have noted that many of the coefficients in an input-output table are not important in the sense that their correct estimation does not significantly influence the results obtained in using the derived input-output model for purposes of impact analysis or forecasting. In fact, West and Jensen found that if they ranked the coefficients by size, they would set the bottom 50 percent to zero before the errors reached 5 percent in the estimation of total output. Drawing on these ideas, Hewings and Romanos (1981) developed a semisurvey technique for the purposes of assembling a set of regional social accounts for a small region in Greece. A new large-scale development project involving both the agriculture and manufacturing sectors was being planned for the region and estimates of the likely impacts on output, employment, and income were needed.

Using the Greek national input-output tables, a survey questionnaire was administered to a sample of firms in each industry in the region. The five largest input and output (sales) coefficients were identified for each industry from the national tables. The reason for the choice of the five largest coefficients was based on empirical evidence showing that, in small tables, the four or five largest coefficients often account for 80 percent to 90 percent of the total requirements from other industries. In the Greek national model, the input coefficient for the fifth largest input in any column was often of the order of 0.01. The regional firms were asked to compare their inputs and outputs with the national averages and to make two adjustments—first to the overall values; and second, to estimate the proportions of inputs and outputs that were derived from or sold to firms within the region. In addition, firms were also asked for information about their total purchases and sales to other industries [to provide estimates of u(r) and v(r)] and the components of value added. Supplementary data were obtained on consumption expenditures and government purchases. Hence, for a relatively small outlay, the larger coefficients in the input-output table were estimated from survey data. The remaining coefficients were adjusted to conform to the survey-derived vectors of total intermediate inputs and outputs. Figure 5.1 provides a comparison for the 22 sectors identified at the regional level of the multipliers estimated using a location quotient technique (modifying the national coefficients) and the semisurvey estimates. These variations hide some of the more important differences in the coefficients—the set of coefficients whose correct estimation is crucial to the accurate estimation of total output. Of course, these are not necessarily the largest coefficients; as West (1983) and Bullard and Sebald (1977) have shown, position in the matrix is also an important consideration. Figure 5.2 shows the degree of correspondence between the three sets of coefficient matrices in terms of this criterion. Many of the

	Multipliers Using Survey Data		
	Output	*Type 1 Income*	*Type 2 Income*
1	1.1230	1.0449	1.3046
2	1.3908	2.5472	3.8116
3	1.0544	1.1033	1.4320
4	1.2736	1.1687	1.5523
5	1.0902	1.0676	1.3433
6	1.0373	1.1805	1.4913
7	1.0451	1.1805	1.6036
8	1.0337	1.0106	1.2411
9	1.0365	1.0460	1.3004
10	1.3139	1.3606	1.7714
11	1.1775	2.0824	3.3689
12	1.1077	1.1110	1.4126
13	1.0864	1.1520	1.5353
14	1.1155	1.1516	1.4492
15	1.2793	3.1678	5.8412
16	1.0265	1.0181	1.2546
17	1.0823	1.0708	1.3333
18	1.2638	1.1488	1.4941
19	1.1744	1.3068	1.7949
20	1.1629	1.1394	1.4298
21	1.0222	–	–
22	1.0569	1.1878	1.5829
	Multipliers Using Location Quotient Adjusted Greek Coefficients		
1	1.0615	1.0424	1.2437
2	1.4031	2.9495	3.5189
3	1.0661	1.1307	1.3490
4	1.1888	1.1482	1.3698
5	1.0406	1.0329	1.2324
6	1.4757	6.3556	7.5827
7	1.2075	2.0515	2.4476
8	1.0582	1.1392	1.3592
9	1.1547	1.6927	2.0195
10	1.1069	1.2266	1.4634
11	1.1004	1.4874	1.7745
12	1.1445	1.2551	1.4974
13	1.0678	1.2288	1.4660
14	1.0415	1.0569	1.2610
15	1.1080	2.2073	2.6335
16	1.1210	1.1856	1.4146
17	1.0849	1.0822	1.2912
18	1.0952	1.0632	1.2684
19	1.1023	1.1678	1.3933
20	1.1707	1.1366	1.3560
21	1.0223	–	–
22	1.0280	1.0835	1.2926

SOURCE: Hewings and Romanos, 1981.

Figure 5.1 Multiplier Comparisons for Evros Region Study

	LQ	*National Table*	*Total Inverse Important*
Survey Table	8	11	18
Location Quotient Table		14	18
Natural Table			38

Figure 5.2 Correspondence Between Inverse Important Coefficients

differences between the national and the semisurvey models may be ascribed to the poor representation of some sectors in the Evros region. These aspects of the regional model are very critical, especially in cases in which the model is used to estimate the impact of a project that might directly effect on a small number of sectors. Figure 5.3 shows a comparison of the estimates impacts of a randomly generated change in final demand on the Evros region using the unadjusted national coefficients, the location quotient-adjusted national coefficients, and the survey-based table. The differences on a sector-by-sector basis are sometimes large. However, note what happens when households are made endogenous; the magnitude of the impacts generated by the inclusion of this one additional sector is enormous. This research and several other subsequent pieces of work have confirmed the overriding importance of consumption-induced linkages at the regional level. The Evros study also revealed that a large percentage of the important coefficients was to be found in the household row and column. However, the a priori estimation of key or important coefficients, in the absence of an input-output table, remains an important area for further research.

ROUND'S TWO-REGION TECHNIQUE

To date, very little has been said about the estimation of other parts of the input-output table at the regional level. The accurate estimation of the components of regional final demand would seem to be just as crucial to the estimation of the regional coefficients. However, with the exception of McMenamin and Haring (1974), little attempt has been made to address this problem. Round (1972, 1979, 1983) has explored this issue in the context of the development of a two-region model. Although the interregional models will form the focus of the next chapter, Round's technique will be discussed here. The technique should be regarded as a semisurvey technique as it does require the estimation of final demand in the two regions (usually one region and the rest of

Sector Number	Change in Final Demand	Evros Survey	Closed Form Output: Location Quotient	Closed Form Output: Greek Coefficients	Closed Form Output: Households Endogenous
1	816.47	939.82	1129.07	1222.84	1072.55
2	309.95	313.80	340.20	353.90	529.77
3	763.93	769.88	783.20	792.31	777.42
4	78.56	78.64	78.94	79.03	109.04
5	61.21	134.74	126.51	372.56	236.48
6	277.56	319.87	278.25	409.20	856.67
7	988.72	1006.86	991.49	1511.96	1168.50
8	188.76	194.40	188.78	199.55	420.42
9	174.53	183.33	174.85	248.30	193.28
10	530.60	616.27	532.12	874.38	713.55
11	709.97	815.98	713.15	1278.84	1217.59
12	496.26	522.01	496.56	557.34	546.95
13	889.74	947.45	891.36	1169.43	1044.83
14	482.37	495.54	482.92	600.13	597.89
15	772.31	873.93	774.56	1192.23	996.25
16	774.52	818.60	806.00	918.48	894.75
17	89.36	97.77	103.53	120.64	125.64
18	3.12	91.60	200.30	271.36	722.39
19	232.16	304.63	368.88	426.25	756.84
20	429.68	737.87	703.32	778.51	818.35
21	91.72	92.68	91.72	91.72	392.11
22	470.70	496.42	571.63	604.14	962.75
23	133.63	–	–	–	3509.41

Figure 5.3 Final Demand and Sectoral Output for Various Evros Models

the country). Round's procedure involves a two-stage estimation for the regional requirements coefficients. Taking the relation that we have developed earlier:

$$_{11}r_{ij} = {}_1t_{ij}a_{ij}(r) \qquad [5.13]$$

where the numerical subscripts refer to the two regions and $_1t$ represents the proportion of the total input supplied from within region 1. A first estimate of $_1t$ may be given by the use of a quotient technique, such as the location quotient described earlier. Round's technique builds on the following relationships:

$$\text{if } {}_1q_{ij} < 1 \text{ then } {}_2q_{ij} > 1 \qquad [5.14]$$

and thus

$$\text{if } {}_1t_{ij} < 1 \text{ then } {}_2t_{ij} = 1 \qquad [5.15]$$

The reasons for these relationships may be derived from the fact that if region 1 has a location quotient greater than 1, it implies that it meets its own needs and exports some of the commodity i to the other region. Concomitantly, region 2 must be an importer of commodity 1. Hence, the relationships implied in equations 5.13 and 5.14. Using these restrictions, Round proceeds to evaluate several second stage techniques that serve to adjust these initial estimates to conform to known vectors of intermediate output. [Because final demand and total output are known, the vector of intermediate output may be derived by subtraction; $u_i(r) = X_i - f_i$]. By adopting a two-region framework, any adjustment in region 1 must be accommodated by a compensating adjustment in region 2. Thus, the procedure at least ensures that total flows within the system sum to known values—a characteristic that cannot be claimed by the use of a quotient technique applied to a single region.

Summary

There are a variety of ways in which input-output tables can be generated from other than survey data. However, there is general agreement that the quality of the tables produced even in concert with only a small amount of survey information are far superior to those that rely exclusively on nonsurvey procedures. Given the complexity of regional economies, it would indeed be surprising if some simple quotient technique was able to yield accurate estimates of the whole regional input-output table. The work of Burford and Katz (1977), who aim to estimate input-output type multipliers without much consideration for the nature of the rest of the input-output table, has generated acceptable estimators. However, their procedures provide little information on the structure of production at the regional level and provide little guidance for ways in which their techniques could be used in contexts (to be described in Chapter 7) in which the input-output model is linked with other regional models.

6. MULTIREGIONAL AND INTERREGIONAL INPUT-OUTPUT MODELS

On several occasions in previous chapters, reference has been made to interregional imports and exports. Their placement in the primary inputs and final demand entries within the regional input-output model suggests that their role in the regional economic growth and development process may be no different than the other components of final demand and primary inputs. However, this may not be the case, if we are able to expand the model in a way in which the interregional interactions are made more explicit. One of the reasons why we want to do this may be ascribed to the presence of *interregional feedback effects*. These effects may occur in the following fashion. Assume that we have two regions, r and s, and that a new activity had been created in region r—for example, a new government installation such as a military base or a research laboratory employing many hundreds of people. The new expenditures in region r will create increased output in that region; this increased output in region r will also necessitate new imports from region s. In order to meet these new import requirements, industries in region s will have to expand their production and thus they may require imports from region r. Hence, output in region r may go up again as a result of the fact that it increased in the first place! These additional demands are known as the feedback effects.

Empirical evidence on the magnitude of feedback effects is scanty. However, Beyers (1976) was able to show that for some industries in the state of Washington, the interregional feedback effects (in a two region, Washington-rest of the U.S. model) were larger than the intraregional indirect effects. Greytak (1970) has also explored some of the empirical aspects of feedback and found them to be significant. The reason for the lack of substantial evidence on this point is the near absence of interregional models for most countries. The cost and complexity involved in their construction have precluded their widespread adoption. Some exceptions are the interregional models for Japan (see Polenske 1970) and the Multiregional Input-Output Model (MRIO) that has been developed for the United States (Polenske 1976) and the commodity-industry interregional input-output model developed for Canada by Hoffman and Kent (1976).

Interregional Input-Output Framework

Interest in interregional modeling was evident early on in the development of input-output models. Both Leontief (1953) and Isard (1953) provided some initial statements and suggestions, some of which were developed by Moses (1955). Miller's work on the magnitude of the interregional feedbacks produced some rather confusing and conflicting results (Miller 1966, 1969); however, it was the work on feedbacks conducted within an expanded economic base-regional income multiplier model developed by Brown (1967) and Steele (1969) that demonstrated the importance of these influences, especially in smaller, very open regional economies. Subsequent research especially by Miller and Blair (1981) and Blair and Miller (1983) on the interaction of spatial and sectoral aggregation issues in interregional and multiregional models has continuied to provide evidence for the need to recast the regional models into at least a two-region context. Thus, even in the development of a model for a single region, it would be preferable to construct a model with two "regions"—the region in question and the rest of the country of which it is a part.

The interregional model will now be presented for the two-region case; essentially, the vector of interregional exports and the vector of interregional imports are expanded into matrices. Accordingly, the regional requirements' matrix is expanded as shown in Figure 6.1. The elements of R_{11} are the requirements from industries in region 1 per unit of output in region 1. A similar definition describes the elements of R_{22} except that the industries are located in region 2. The remaining matrices describe the interregional transactions. The elements of R_{12} may be interpreted in two ways; they are the export flows from region 1's industries to those in region 2 or they may be interpreted as the imports from industries in region 1 required per unit of output in region 2. The elements of R_{21}, similarly, are imports from region 2 by industries in region 1 or exports from region 2 to 1. A hypothetical two-region transactions matrix is shown in Figure 6.2; the corresponding equation system is now

$$\left[\begin{array}{c:c} R_{11} & R_{12} \\ \hdashline R_{21} & R_{22} \end{array}\right] \begin{bmatrix} X_1 \\ \cdots \\ X_2 \end{bmatrix} + \begin{bmatrix} Y_1 \\ \cdots \\ Y_2 \end{bmatrix} = \begin{bmatrix} X_1 \\ \cdots \\ X_2 \end{bmatrix} \qquad [6.1]$$

$$R = \begin{bmatrix} R_{11} & \vdots & R_{12} \\ \cdots & \vdots & \cdots \\ R_{21} & \vdots & R_{22} \end{bmatrix}$$

Figure 6.1 Two-Region Interregional Model

Note that the vectors of final demand (Y) and total output (X) have been divided into the two region components. The solution of equation 6.1 is:

$$\begin{bmatrix} B_{11} & \vdots & B_{12} \\ \cdots & \vdots & \cdots \\ B_{21} & \vdots & B_{22} \end{bmatrix} \begin{bmatrix} Y_1 \\ \cdots \\ Y_2 \end{bmatrix} = \begin{bmatrix} X_1 \\ \cdots \\ X_2 \end{bmatrix} \quad [6.2]$$

where the submatrices B_{11}, B_{12}, B_{21}, and B_{22} are the partitioned components of the Leontief inverse matrix of the solution to equation 6.1. If we produce the solution for X_1 we have:

$$X_1 = B_{11}Y_1 + B_{12}Y_2 \quad [6.3]$$

$$= [(I - R_{11}) - R_{12}(I - R_{22})^{-1}R_{21}]^{-1}Y_1$$

$$+ [(I - R_{11}) - R_{12}(I - R_{22})^{-1}R_{21}]^{-1}R_{12}(I - R_{22})^{-1}Y_2 \quad [6.4]$$

Here we have the two components of final demand, Y_1 and Y_2, creating direct and indirect responses in output in region 1. Let us examine the term inside the square brackets. In addition to the intra-regional coefficients for region 1 (R_{11}), we have the expression $R_{12}(I - R_{22})^{-1}R_{21}$. The term R_{21} contains the demands from industries in region 2 necessary to produce output in region 1; these demands will stimulate output in region 2, directly and indirectly $(I - R_{22})^{-1}$, and these demands in their turn will require inputs from region 1, R_{12} (what we referred to earlier as the interregional feedback effects). The expression $R_{12}(I - R_{22})^{-1}Y_2$ translates final demand in the second region into total output in the second region and then into input requirements from region 1. Hence, total output in region 1 may be responsive to demands initially placed in region 2 as well as to those placed directly in region 1. As the number of regions increases, the interregional effects would become much more complex; for purposes of exposition, we will retain

From		To Region 1: 1	2	3	4	5	To Region 2: 1	2	3	4	5	Final Demand*	Total Output
	1	21	–	9	3	–	10	–	5	5	–	47	100
	2	1	8	7	29	–	–	–	10	5	–	40	100
Region 1	3	3	20	–	50	7	–	–	–	3	–	17	100
	4	31	2	38	–	3	–	–	1	–	–	25	100
	5	10	25	26	1	4	–	4	–	–	–	30	100
	1	2	–	–	–	–	17	1	8	–	–	87	120
	2	–	–	–	–	–	1	6	1	31	–	37	80
Region 2	3	–	4	–	–	15	4	16	–	57	45	34	150
	4	–	–	1	–	6	36	1	60	–	19	4	120
	5	5	–	3	–	9	18	20	45	1	28	68	210

*Excluding interregional exports.

Figure 6.2 Interindustry Transactions Matrix: Two-Region Case

the two-region example.[8] Using the data in Figure 6.2, equation 6.3 was solved for X_1; it was assumed that the change in final demand in each sector in region 1 was $10 million whereas there was no change in final demand in any sector in region 2. Hence, the second term of equation 6.4 reduces to zero. For purposes of comparison, the equation system was also solved under an assumption of no feedback effects, namely,

$$\Delta X = (I - R_{11})^{-1} \Delta Y_1 \qquad [6.5]$$

The results are shown in Figure 6.3; the feedback effects range from 3 percent to 7 percent of the total effects. Although these are not very large, they still represent effects that could translate into several hundred additional jobs (assuming that the employment output figures shown in Figure 4.2 are correct, the feedback effect in sector 1 of $1.179 would generate about 236 extra jobs). However, if the effects of consumer expenditures had been included in the model, the feedback effects might have been much larger (see Airov, 1967 for an interregional formulation of a Keynesian multiplier model).

The Multiregional Input-Output Model

There have been a number of contributions to this formulation; however, the model developed by Leontief and Strout (1963) and some subsequent modifications made by Polenske (1972) have become the most widely known. The Leontief-Strout formulation has been recast

	(1) Regional Model Equation (3.32)	(2) Interregional Model Equation (3.31)	(3) Feedback (2) − (1)	(4) Feedback as Percentage of (1)
1	17.277	18.456	1.179	6.824
2	22.378	23.946	1.568	7.006
3	31.074	32.139	1.065	3.427
4	28.414	29.321	0.907	3.192
5	26.756	27.704	0.948	3.543

Figure 6.3 Regional and Interregional Models Compared: Change in Total Output Associated with $10 Million Change in Final Demand in Each Sector

into the entropy maximization framework by Wilson (1970) providing for a direct link between commodity flow modeling, input-output, linear programming, and the entropy paradigm (see Kim et al. 1983 for a recent application of such a model to Korea). A summary of the Korean application will be provided in the next section.

Figure 6.4 shows the reduced form solutions for three multiregional models, the Leontief-Strout gravity model and the row coefficient and column coefficient models developed by Polenske (1970). In the interregional input-output model, requirements are identified on an industry- and region-specific basis, such that a typical element might be ${}_{pq}r_{ij}$, the requirements from industry i in region p per unit of output of industry j in region q. In the multiregional model, flows between regions are assumed to move into "pools" from which intraregional firms draw for their requirements. These pools may be thought of as collections of the output of the various sectors, differentiated by product but not by region of origin. Thus, in the Leontief-Strout formulation, firms are assumed to be indifferent to the origin of their purchases. Thus, the interregional flows, ${}_{pq}X_i$, show only the identity of the industrial sector moving the flow between region p and q, not the ultimate destination on a sector-specific basis within region q.

The general form of the Leontief-Strout model is:

$$ {}_{pq}X_i = \{[{}_{ro}X_i \cdot {}_{os}X_i] / {}_{oo}X_i\}\ {}_{pq}K_i \qquad [6.6] $$

where ${}_{pq}X_i$ is the supply total shipment of good i from the supply pool in region p to the demand pool in region q, ${}_{ro}X_i$ is the supply pool in region r, ${}_{os}X_i$ is the demand pool in region s, ${}_{oo}X_i$ is the total amount of i produced in the nation, and ${}_{pq}K_i$ is an empirical constant.

	Row Model	Column Model	Gravity Model
Trade Coefficient Equation	$_{pq}X_i = {}_{pq}R_{i\,po}X_i$	$_{pq}X_i = {}_{pq}C_{\,oq}X_i$	$_{pq}X_i = \dfrac{[_{po}X_{i\,oq}X_i]}{_{oo}X_i} \cdot {}_{pq}K_i$
In Matrix Form	$R'\Delta X = A\Delta X + \Delta Y$ $\Delta X = [R' - A]\Delta Y$	$\Delta X = C(A\Delta X + \Delta Y)$ $\Delta X = (I - CA)^{-1}C\Delta Y$	$T'\Delta X = S(A\Delta X + \Delta Y)$ $\Delta X = (T' - SA)^{-1}S\Delta Y$

In the gravity model

$$_{pq}S_i = \begin{cases} {}_{po}X_i \left[1 - \dfrac{_{oo}X\,{}_{pq}X_i}{_{po}X_{i\,oq}X_i} \right] & \text{when } p \neq q \\ 1 & p = q \end{cases}$$

$$_{pq}T_i = \begin{cases} {}_{po}X_i \left[1 - \dfrac{_{oo}X\,{}_{qp}X_i}{_{po}X_{i\,oq}X_i} \right] & \text{when } p \neq q \\ 1 & p = q \end{cases}$$

SOURCE: Polenske, 1970.

Figure 6.4 Gravity, Row, and Column Multiregional Models

The relationship between this formulation and the gravity model (see Haynes and Fotheringham, 1984) should be clear. Equation 6.6 contains an expression that regards flows as a function of (1) supply and demand in the region of origin and destination, respectively, (2) total production of that commodity in the whole system and, (3) as a function of some constant. This latter term is developed further to include some measure of the physical distance separating the two regions:

$$_{pq}K_i = (_rC_i + {}_sH_i)_{pq}d_i \qquad [6.7]$$

where C and H are empirical constants and d is a measure of the reciprocal of distance between p and q for commodity i. This can be adjusted to a value of physical distance, time, or costs and thus may vary depending on the nature of commodity or industrial output i.

Figure 6.4 shows the matrix formulations for the Leontief-Strout and the two other models developed by Polenske. In the only empirical application in which the models were tested against survey data, using data for the Japanese economy, Polenkse (1970) found that although the gravity model performed the best of all three, the column coefficient model performed almost as well.

The Interregional Model in an Entropy Formulation

Haynes and Fotheringham (1984), in a companion monograph in this series, describe a number of spatial interaction models. A careful reading of Chapter 2 of that monograph will assist the understanding of the final section of this chapter in which we will explore the links between these models and the input-output system and report on an application to Korea. Wilson (1970) derived four types of linkage between input-output and commodity flow models; we will develop the one in which regional production and consumption are given. The objective here is to find an efficient allocation of production and consumption, which we will assume will be accomplished if total transportation costs are minimized subject to a set of constraints:

$$\text{Min } C_h = \sum_p \sum_q {}_{pq}C_i \, {}_{pq}X_i \qquad [6.8]$$

$$\text{Subject to:} \quad \sum_q {}_{pq}X_i = \sum_j {}_p r_{ij} \sum_q {}_{pq}X_i + {}_pY_i \qquad [6.9]$$

$$\sum_q {}_{pq}X_i = {}_pX_i \qquad [6.10]$$

$$\sum {}_{pq}X_i = {}_qY_i \qquad [6.11]$$

$$S_i = -\sum_p \sum_q {}_{pq}X_i \ln({}_{pq}X_i) \qquad [6.12]$$

$${}_{pq}X_i \geqslant 0 \text{ for all p and q} \qquad [6.13]$$

Equation 6.8 is merely a statement of the objective, which is to minimize total transportation costs. Constraint 6.9 is a restatement of the accounting balance of equation 3.14 with the differences that some of the demands placed on the output for industry i in region p come from all other regions (hence the summation over all q regions). Equations 6.10 and 6.11 ensure that total demand and total supply for commodity i are in balance over the whole system of regions. The inclusion of the term 6.12 provides a scalar measure of the dispersion of flows in the

Commodity Group	β	R^2 Regression, Estimated Against Observed
1	1.8330	0.68
2	1.9444	0.57
3	1.0555	0.25
4	2.0277	0.50

SOURCE: Kim et al., 1983.

Figure 6.5 Calibration Results: Koizan Interregional Model

economy (see Erlander 1980); it avoids the linear programming formulation restricting flows of a commodity to one direction between two destinations. The phenomenon of two-way flows of commodities, known as cross-hauling, has been observed for many commodities. It reflects differences in taste, product differentiation, and so forth. The general solution for this model yields:

$$ {}_{pq}X_i = {}_pA_i \cdot {}_pX_i \cdot {}_qB_i \cdot {}_qY_i \exp(-\beta_i \cdot {}_{pq}C_i) \qquad [6.14] $$

This formulation is related to equation 6.6; the major difference lies in the inclusion of A and B terms, which ensure that all flows are located from regions to meet the constraint set 6.10 and 6.11.

The model presented above, with some additional embellishments (see Kim et al. 1983) was applied to four groups of commodities in the Korean economy for 1978. The groups were characterized as (1) primary products such as coal or limestone; (2) cement and nonmetallic minerals; (3) agricultural products, and (4) manufactured goods. Given the high degree of aggregation, the results provided in Figure 6.5 were rather encouraging for groups 1, 2, and 4. The explained variation in flows was between 50 percent and 70 percent; commodity group 3 was modeled less successfully; only 25 percent of the variation was explained by the model. Part of the reason for the high level of unexplained variation may be attributed to aggregation, poorly developed transportation systems, and the fact that agricultural products in Korea tend to move over relatively short distances. This may be ascertained by the fact that the value of β is much higher, indicating a greater degree of sensitivity to the value of β is much higher, indicating a greater degree of sensitivity to transportation costs.

Summary

The success involved in linking input-output and commodity flow models provides one of several examples of the benefits from integrating models. In the next chapter, some additional experiments in linking input-output models with other modeling ventures will be presented. However, it should be stated that interregional input-output modeling presents many additional problems not faced by single-region analysis, not the least of which is the daunting issue of data collection. The issues of coefficient stability raised in the context of single-region modeling are joined by problems relating to the stability of the *trading* relations between regions.

7. EXTENSIONS TO INPUT-OUTPUT ANALYSIS

One of the most promising developments in regional and interregional input-output analysis in recent years has not been the further development of the input-output model per se, but rather the linkage of input-output analysis within much more extensive modeling systems. In Chapter 2, the link between economic base, Keynesian macro accounting analysis, and input-output systems was demonstrated. This chapter will begin by developing a link with another useful framework, the Kalecki system as interpreted in input-output terms by Miyazawa (1976). This extension was used as the basis for the development of much of the demometric/demographic-economic analysis associated with the work of Schinnar (1976, 1977) and Batey and Madden (1983) and is consistent with a parallel development in what may be referred to as social accounting systems (SAMs). Miyazawa was interested in the incorporation of the distribution of income and its expenditure into the input-output framework. The idea will be developed first with a simple one-sector model, and thereafter it will be extended to the more general case.

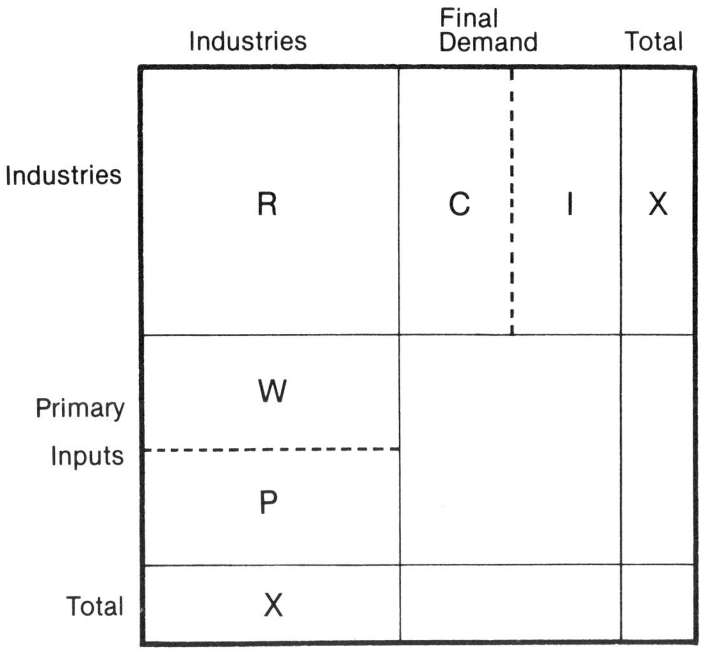

Figure 7.1 Miyazawa Framework

Miyazawa's Framework

The link between the structure of production and the distribution of income is an intriguing one, particularly in the development context. The choice of any particular project, such as a new irrigation system, a new highway network, or a large-scale iron and steel plant is likely to change the structure of production in a very open regional economy. In so doing, it will possibly change the distribution of income; the expenditures resulting from these income changes may themselves create further effects on the production system and so on. In this and the next section, two proposals for handling these links will be described and illustrated using data for Sri Lanka (Pyatt and Roe 1977). Although not a regional economy, it has many of the characteristics of one, and the availability of excellent data for this country provides an unusual opportunity to examine production-income interactions in some detail. Figure 7.1 shows the standard framework; there is one intermediate sector, R; two components to final demand, consumption, C, and investment, I; and two primary inputs, W (wages) and P (profits). Hence, the accounting system is:

$$R + C + I = X \quad [7.1]$$

$$R + W + P = X \quad [7.2]$$

If the usual conventions are followed, then we have:

$$r = R/X \quad \text{and} \quad f = C + I; \text{ then}$$

$$X = (1 - r)^{-1}f \quad [7.3]$$

Assume that C = cY and hence = c(W + P); then we have the Keynesian solution:

$$Y = (1 - c)^{-1}I \quad [7.4]$$

Given that C + I = W + P (i.e., f = Y), by substitution of equation 7.4 into equation 7.3:

$$X = (1 - r);\ f = (1 - r)^{-1}(1 - c)^{-1}I \quad [7.5]$$

This holds only for a particular income distribution; if the shares of wages and profits, d_1 and d_2 are, respectively:

$$d_1 = W/Y; \quad d_2 = P/Y \quad [7.6]$$

and consumption is similarly disaggregated into the share by wage-earners and the share by those receiving profits:

$$C = C_1 + C_2 = c_w/W + c_p/P \quad [7.7]$$

The generalized income multiplier may now be written as:

$$(1 - c)^{-1} = [1 - (c_1d_1 + c_2d_2)]^{-1} \quad [7.8]$$

If the value added ratio, v = Y/X = (1 – r) is further subdivided into: $v_1 = W/X$ and $v_2 = P/X$, we now have:

$$X = (1 - a)^{-1}[1 - (c_1v_1 + c_2v_2)\ (1 - r)^{-1}]I \quad [7.9]$$

This scalar system may now be expanded by dividing the input-output part into n sectors and k income groups. Hence, we now have:

R, an $n \times n$ matrix of regional input coefficients;

V, a $k \times n$ matrix of value added ratios (for wages and salaries);

C, an $n \times k$ matrix of consumption coefficients (out of wages and salaries);

f_c, an $n \times 1$ vector of consumption demand; and

f', an $n \times 1$ vector of the rest of final demand.

The accounting balance is now:

$$X = RX + f_c + f' \qquad [7.10]$$

for which the normal solution would be of the form:

$$X = [I - R]^{-1}[f_c + f'] \qquad [7.11]$$

Instead, the consumption effect will be made endogenous in a way that will enable the analyst to measure the effects of changes in consumption without repeated matrix inversion. The matrix f_c may be defined as CVX because VX represents the income accruing to consumers from wages and salaries and C represents the expenditure pattern per unit of income (i.e., average propensities to consume). Hence, equation 7.10 may be written:

$$X = RX + CVX + f' \qquad [7.12]$$

which yields the closed form solution:

$$X = [I - A - CV]^{-1}f' \qquad [7.13]$$

If B is defined as the usual Leontief Inverse matrix (i.e., $[I - A]^{-1}$), then we may write:

$$X = B[I - CVB]^{-1}f' \qquad [7.14]$$

$$X = B[I + CKVB]^{-1}f' \qquad [7.15]$$

where $K = [I - L]^{-1}$ and $L = VBC$. L is the macroeconomic propensity to consume; with only one income group, K would reduce to a scalar. KVB may be regarded as the multisector income multiplier. With this formulation, we are now ready to explore the structure of the propagation process:

$$\begin{bmatrix} X \\ \cdots \\ Y \end{bmatrix} = \begin{bmatrix} A & \vdots & C \\ \cdots & \vdots & \cdots \\ V & \vdots & 0 \end{bmatrix} \begin{bmatrix} X \\ \cdots \\ Y \end{bmatrix} + \begin{bmatrix} f' \\ \cdots \\ g \end{bmatrix} \qquad [7.16]$$

where g is exogenous income. Further manipulation reveals:

$$\begin{bmatrix} X \\ \cdots \\ Y \end{bmatrix} = \begin{bmatrix} I-A & \vdots & -C \\ \cdots & \vdots & \cdots \\ -V & \vdots & I \end{bmatrix}^{-1} \begin{bmatrix} f' \\ \cdots \\ g \end{bmatrix} \qquad [7.17]$$

$$= \begin{bmatrix} B[I+CKVB] & \vdots & BCK \\ \cdots & \vdots & \cdots \\ KVB & \vdots & K \end{bmatrix} \begin{bmatrix} f' \\ \cdots \\ g \end{bmatrix} \qquad [7.18]$$

In an analysis of this modeling framework applied to a three-region case study for Japan, Miyazawa (1976) expanded the interpretation of L such that a typical element 1^{pq} shows how much income in region p is generated by an additional unit of income in region q. He was able to show the degree to which the middle region of Japan (including the Tokyo region) had an 83.5 percent self-sufficiency ratio for income, whereas the northeast and western regions were dependent on the middle for 40 percent and 35 percent, respectively, of their incomes. This formulation presented by Miyazawa anticipates a great deal of the conceptual development of social accounting matrices (a topic to be discussed below). Prior to that discussion, an attempt was made, using some data for Sri Lanka (Pyatt and Roe 1977) to develop a Miyazawa system of equations. In this case, twelve industrial sectors and three income groups (urban, rural, and estate) are identified. The partitioned solution for equation 7.18 is shown in Figure 7.2; the bottom righthand 3×3 matrix is K, the matrix showing the direct and indirect effects on income changes on each income group. Of course, the Miyazawa framework is not limited to applications focusing on interregional distribution effects; we may note several attempts to expand the framework to account for the *distributional* effects of projects—namely, their effects on different income groups (see Golladay and Haveman 1976

	1	2	3	4	5	6	7	8	9	10	11	12	1	2	3
	Tea	*Rubber*	*Coconut*	*Paddy*	*Other Agri-culture*	*Agri-cultural Processing*	*Mining*	*Tra-ditional Industry*	*Modern Industry*	*Con-struction*	*Trade Transport*	*Services*	*Urban Labor*	*Rural Labor*	*Estate Labor*
Industrial Sectors															
1	1.0069	.0049	.0021	.0073	.0028	.0047	.0019	.0031	.0014	.0016	.0030	.0035	.0058	.0102	.0094
2	.0008	1.0004	.0002	.0005	.0003	.0004	.0002	.0067	.0003	.0005	.0004	.0002	.0000	.0000	.0000
3	.0541	.0353	1.0140	.0457	.0188	.1542	.0121	.0389	.0098	.0116	.0197	.0183	.0237	.0375	.0463
4	.0969	.0638	.0260	1.1508	.0375	.5994	.0229	.1196	.0217	.0249	.0378	.0423	.0000	.0000	.0000
5	.1900	.1183	.0469	.1503	1.1155	.1153	.0599	.1249	.0366	.0439	.0701	.0804	.0001	.0002	.0002
6	.1807	.1191	.0486	.1588	.0700	1.1216	.0426	.2236	.0372	.0462	.0704	.0789	.0000	.0001	.0001
7	.0008	.0006	.0002	.0006	.0003	.0005	1.0001	.0020	.0027	.0429	.0006	.0021	.0000	.0000	.0000
8	.1293	.0724	.0292	.0923	.0497	.0723	.0286	1.1245	.0486	.0904	.0616	.0405	.0002	.0003	.0003
9	.2123	.1441	.0473	.1289	.0730	.1006	.1347	.0918	1.2993	.1370	.0989	.0638	.0012	.0021	.0022
10	.0043	.0031	.0017	.0046	.0019	.0033	.0115	.0026	.0016	1.0945	.0063	.0333	.0000	.0000	.0000
11	.2902	.1789	.0766	.1993	.0947	.1954	.1206	.1987	.1652	.1994	1.1430	.1179	.0009	.0015	.0017
12	.1026	.0779	.0447	.1235	.0496	.0839	.0318	.0599	.0313	.0351	.0662	1.0568	.0003	.0005	.0006
Income Groups															
1	.1105	.0718	.0457	.1004	.1000	.0927	.0403	.1431	.1065	.0841	.1991	.1672	1.0863	.0979	.1029
2	.2532	.3500	.2015	.8344	.2800	.5103	.2108	.3045	.1156	.1495	.2520	.1853	.1695	1.2176	.2409
3	.6146	.2522	.0411	.0098	.0103	.0167	.0031	.0083	.0027	.0055	.0096	.0118	.0095	.0138	1.0141

NOTE: Rows and columns 1-12 are industrial sectors.

Figure 7.2 Empirical Implementation of Equation 7.18 Using Data from Sri Lanka

Sector	*Change in Final Demand*	*Change in Output*	*Change in Final Demand*	*Change in Output*
1	20	20.187	0	.107
2	10	10.020	0	.011
3	0	1.434	0	3.182
4	0	2.575	0	12.204
5	0	4.984	0	2.671
6	0	4.806	20	22.803
7	0	.022	0	.036
8	0	3.310	0	1.931
9	0	5.688	10	15.004
10	0	.116	0	.083
11	0	7.593	0	5.561
12	0	2.830	0	1.990
Total	30	63.565	30	65.583

Class of Labor	*Income Effects of Change in Final Demand*		
Urban	1	2.928	2.920
Rural	2	8.564	11.361
Estate	3	14.814	.361

Figure 7.3 Output Effects of Change in Final Demand

and Rose et al. 1982). Over time, one might be able to consider using the model to examine possible "trickle-down" effects of projects—the longer-run impacts that filter down through the economic system from one sector or income group to another. However, most regional and interregional input-output models have not been used very extensively in the context of monitoring change in regional economies.

To illustrate the workings of this system, two projects were assumed to be candidates for evaluation as to their impacts on output and income. The first project was assumed to yield increases in final demand in sectors 1 and 2 only—thus, an agriculturally oriented undertaking. The second project focused on agricultural processing and modern industry. The relevant data are shown in Figure 7.3. In both cases, total additional deliveries to final demand were set at 30. As one might have expected, the distribution of the direct and indirect effects (including endogenously generated income effects, because we are solving 7.18) vary substantially by sector (see Figure 7.3). Note, however, that sector 9 is the beneficiary of an additional output of 5.6 in the first project even though the initial change in final demand in this sector was zero. The role of trade and transport is clearly seen as the sector picks up sizable output effects from both projects. The major differences are to be seen in the income distribution impacts; project 1 favors the second and third groups overwhelmingly, whereas the second project's impacts are almost exclusively concentrated in the rural income groups. However, as we shall see, these are just parts of the social accounting system that will be described below.

The General Social Accounting Framework

The inclusion of an exogenous income term, g, in the Miyazawa framework has been further modified in the social accounting system. Originally pioneered by Stone (1961), these systems have been extended by Pyatt and Roe (1977) and applied at the regional and interregional level by Pyatt and Round (1983), Round (1984), and Bell et al. (1982). Figure 7.4 shows a typical social accounting system for a single region. Note that the system is divided into three major components: factors of production, institutions, and activities. The input-output model is located in the latter group. Wages and salaries paid to workers in the various income groups are transferred to institutions (urban households, rural households, etc.) and the intrainstitutions submatrix enables transactions to take place between these institutions. Here, for example, we would find taxes and transfers between households and the government and various other nonwage and salary transfers. With the increasing importance of these latter forms of income, the social accounting system provides for a more complete picture of the various flows within an economy. The analytical solution may be derived in several ways; first, a generalized inverse may be obtained and the solution vector found in the usual way:

$$X = [I - A^*]^{-1}f' \qquad [7.19]$$

where A* is the matrix of coefficients derived from Figure 7.4. Alternatively, the partitioned matrix may be employed separately and the use of permutation matrices will yield a solution of the form:

$$X = M_1M_2M_3f' \qquad [7.20]$$

In this case, f′ represents final demand outside the region as consumption and government activities are now endogenous. The reader is

		FACTORS OF PRODUCTION	INSTITUTION CURRENT ACCOUNT	PRODUCTION ACTIVITY	INTERREGIONAL & FOREIGN ACCOUNTS
FACTORS OF PRODUCTION	Labor • • • Capital			(Value added in production)	(Net factor incomes from other regions and the rest of the world)
INSTITUTIONS	Firms • Households • Government	(Gross Profits) (Wages, Salaries)	(Transfers and taxes)		(Transfers from outside regions)
PRODUCTION ACTIVITY	Agriculture • • Manufacturing • Services		(Consumer expenditures on regional goods)	(Interindustry transactions)	(Exports of goods outside the region)
INTERREGIONAL & FOREIGN ACCOUNTS			(Consumer expenditures on imports)	(Imports of intermediate goods)	

SOURCE: Pyatt and Roe, 1977.
NOTE: Entries in the matrix provide examples of some of the transactions between accounts.

Figure 7.4 Single Regional Social Accounting Scheme

referred to Pyatt and Roe (1977) for an explanation of the second solution system. Round (1984) has recently proposed a methodology for developing interregional solutions of this type, although the procedures are not quite as tractable once the number of regions exceeds two.

Perhaps one of the most creative extensions of this system has been the work of Bell et al. (1982). They applied the system to the Muda region of Malaysia to evaluate the effects of a large-scale irrigation project. In so doing, they adopted a variation of Tinbergen's semi-input-output method and linked the system to a cost-benefit framework in a similar fashion to the one described in Chapter 4.

Bell et al. (1982) adopted a more extensive analysis in the case of the Muda region irrigation project. The standard balance equation was given as:

$$X_i = \sum_j r_{ij} X_j + \sum_k b_{ik} + V_i + E_i \qquad [7.21]$$

where the second term on the righthand side refers to consumption by the set of household sectors and V and E refer to investment and exports. The income of each household is assumed to be made up of wages and salaries and distributed profits and net transfers:

$$b_k = \Sigma w_{kj} X_j + Tr_k \qquad [7.22]$$

Taxes and savings are similarly defined as linear functions of income and, in the case of taxes, an exogenous component, T*:

$$Tx_k = t_k Y_k + T_k^* \qquad [7.23]$$

$$S_k = s_k (Y_k - T_k) \qquad [7.24]$$

Finally, a linear consumption function is defined as:

$$b_{ik} = \alpha_{ik} + \beta_{ik} B_k \qquad [7.25]$$

where B_k represent total outlays on consumption. In matrix form the system may be written:

$$\begin{bmatrix} I-R & \vdots & -BC \\ \cdots & \vdots & \cdots \\ -\Omega & \vdots & I \end{bmatrix} \begin{bmatrix} X \\ \cdots \\ Y \end{bmatrix} = \begin{bmatrix} \Gamma e & \vdots & B[I-S] \\ \cdots & \vdots & \cdots \\ 0 & \vdots & R^* \end{bmatrix} + \begin{bmatrix} V \\ \cdots \\ 0 \end{bmatrix} + \begin{bmatrix} E \\ \cdots \\ 0 \end{bmatrix} \qquad [7.26]$$

where

B is an $n \times h$ matrix of β_{ik}'s,

C is an $h \times h$ diagonal matrix of c_k's $= (1 - s_k)(1 - t_k)$,

Γ is an $n \times h$ matrix of α_{ik}'s,

E is an $n \times 1$ vector of E_i's,

e is an $h \times 1$ vector of 1's,

V is an $n \times 1$ vector of V_i's,

T^* is an $n \times 1$ vector of T^*_k's, and

Ω is an $H \times n$ matrix of w_{kj}'s.

In this form, equation 7.26 represents the standard regional version of the Leontief models. As with the Miyazawa version described earlier, given the levels of exogenous demand, the model may be solved for outputs and incomes. In their application to the Muda region in Malaysia, Bell et al. (1982) adopted a modified version of the semi-input-output approach proposed by Tinbergen (1966) and further elaborated in Kuyvenhoven (1978). Sectors were classified into tradeable and nontradeable (in a spatial sense), the latter being restricted to transactions within the region. Then, the output of traded goods was assumed to be given, whereas their associated exports were determined endogenously, and the output of the nontraded sectors was assumed to be determined endogenously—as was the level of exports in those sectors, namely zero (by definition). This model—and variations on it—was used to measure the impact of a large irrigation scheme on the region through comparison of regional growth and income distributions with and without the project.

One major difficulty, of course, was that the analysts were confined to the measurement of direct and indirect effects within the region. As we noted in Chapter 6, Round (1972, 1978a, 1978b, 1979, 1983) has argued persuasively for the development of at least two-region models (the region in question and the rest of the country of which it is a part) on the basis of a number of factors—especially the provision of consistency in estimation of the original parameters. Recently, Round (1984) has provided some guidance to the development of two and larger region social accounting systems, incorporating the multiplier decomposition ideas noted in equation 7.20. In the Muda region case, the need for consideration of the extraregional effects became more pressing when rural incomes began to rise above a subsistence level, creating the opportunities for rural households to engage in consumption of additional goods and services. Because many of these would be produced outside the region, the effects of increasing rural incomes might not continue to narrow disparities between rural and urban incomes. Clearly, the changes in rural marginal propensities to consume in total and the changes in consumption of local versus nonlocal goods would play a very important role.

The Batey-Madden Activity Analysis Framework

Two important extensions of the present work at the regional level have been developed by Schinnar (1976) and in various papers by Batey and Madden (see 1983 for references to earlier work). Unlike the attempts by Ledent (1978) to link demographic and economic base models, these authors have linked demographic components with the interindustry model. The work by Batey and Madden has been developed the most extensively to date; drawing on some earlier notions by Tiebout (1969) concerning the role of different labor groups (inmigrants and local residents) in the generation of expenditure effects and some empirical work by Blackwell (1978), who suggested the incorporation of the impacts of previously unemployed residents on local income generation, Batey and Madden have produced some extensions to the Miyazawa framework. Figure 7.5 shows one such extension; in this case, an activity analysis framework is employed in which consumption effects are differentiated by status in the work force (employed or unemployed). In matrix form, this yields:

$$\begin{bmatrix} I-R & -Q \\ -L & I-Z \end{bmatrix} \begin{bmatrix} X_I \\ h \end{bmatrix} = \begin{bmatrix} f_1 \\ f_2 \end{bmatrix} \qquad [7.27]$$

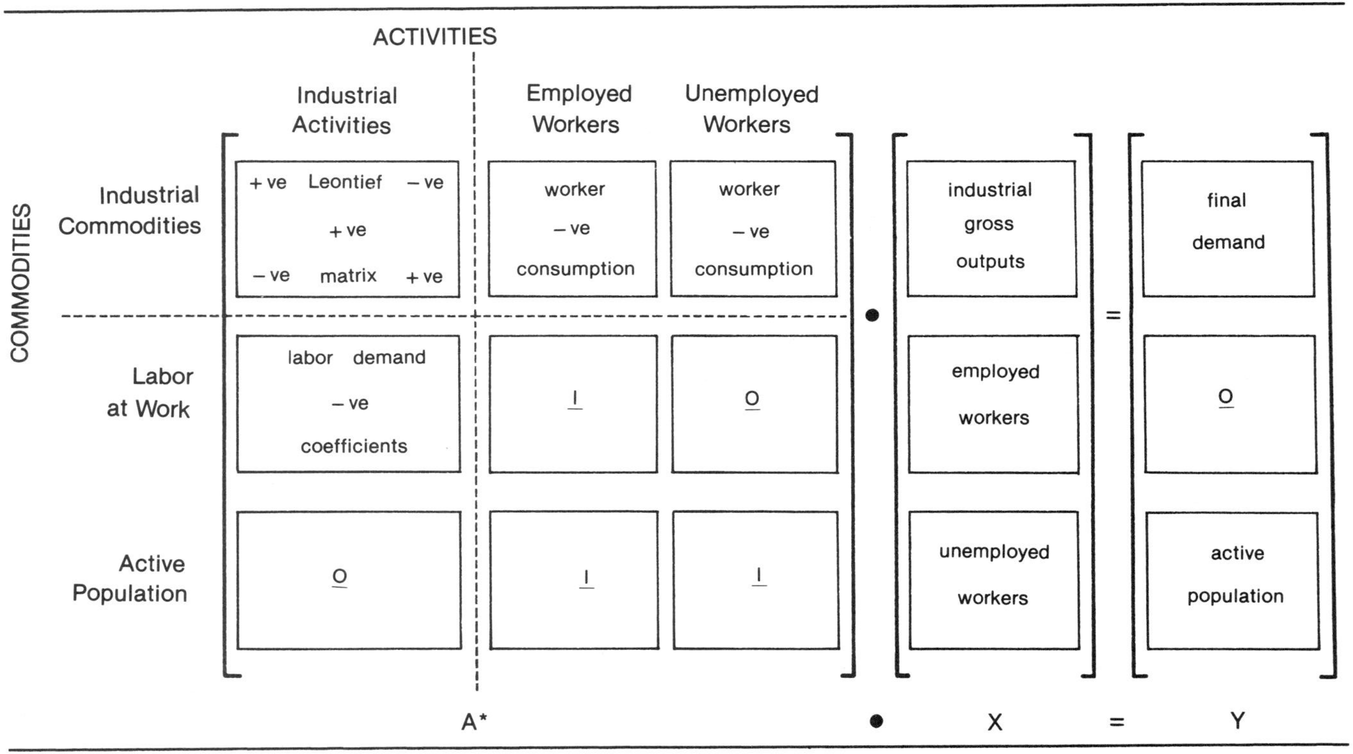

Figure 7.5 The Merseyside Model

where h refers to different types of workers and f_1 and f_2 are the components of final demand. The solution for h is:

$$h = [I - Z - L \cdot (I - R)^{-1} \cdot Q]^{-1}[f_2 + L \cdot (I - R)^{-1} \cdot f_1] \qquad [7.28]$$

In this formulation, it is possible to trace the impacts of changes in final demand on the labor supply. In an empirical analysis of the model in the Merseyside region of the UK, they were able to show that, ceteris paribus, the overall unemployment rate would have been 43 percent higher were it not for the expenditure effects generated by unemployment compensation.

Links with the Labor Market

The role of differences in consumption patterns by different groups within a region has been addressed by van Dijk, Oosterhaven, and Stoffelsma (1984) and Folmer and Oosterhaven (1984) and in the spirit of the Bell et al. (1982) work noted earlier by Oosterhaven (1981). The Oosterhaven work is notable for the fact that it linked an interregional input-output model to traditional cost-benefit appraisal to measure the effect of alternative land reclamation schemes in a small part of Holland. If an accurate attempt to link the role of households to the economic system is to be undertaken, then some careful consideration will have to be made of labor market behavior. Although Batey and Madden (1983) accounted for movements into and out of the employed ranks, van Dijk et al. (1984) examined the regional impact of inmigration and linked it with both the input-output model and a labor force vacancy chain model. The consumption effects of different households inmigrating into the region were considered. In the analysis that follows, it will be assumed that the regional system experiences five changes in demand, only one of which is exogenously determined—namely the consumption demand of inmigrants. The other changes in demand are the changes in regional intermediate demand and consumption-induced effects. These are shown in equation 7.29:

$$\Delta x = A\Delta x + q^i\Delta c^i + q^u\Delta c^u + q^n\Delta c^n + \Delta f' \qquad [7.29]$$

where Δx is the change in production levels, q^i is the regional consumption vector associated with people with high labor incomes, q^u is the consumption of those receiving unemployment benefits, and q^n is the consumption of those with non-labor-force active benefits. The Δc

terms are the associated changes in consumption and $\Delta f'$ the inmigrants' consumption.

The vacancy chain model provides a link between employment creation and changes in output (which are created by the expenditures of the inmigrants) and increases in labor productivity:

$$\Delta e = \ell \Delta x - p^{\ell} e_{t-1} \qquad [7.30]$$

where

Δe is the vector of employment changes by industry,

ℓ is the vector of marginal employment coefficients,

p^{ℓ} is a diagonal matrix of labor productivity increases, and

e_{t-1} is the base year employment levels by industry.

The total number of vacancies is shown below:

$$v = Tv + \Delta e + v' \qquad [7.31]$$

where v is the total number of vacancies and T the interindustry transition probability matrix of people leaving industry i to take up jobs in industry j. v' is a miscellaneous category constituting movements out of the region, out of the labor force because of retirement, and as a result of frictional vacancies. In addition to transfers among industries by those already employed, other vacancies are filled by unemployed who enjoyed benefits (equation 7.32), nonactives who enjoyed nonactive benefits (equation 7.33) or no benefits (equation 7.34), and inmigrants from other regions (equation 7.35):

$$\Delta u = t^{u} v \qquad [7.32]$$

$$\Delta n = (t^{n})' v \qquad [7.33]$$

$$\Delta r = (r^{r})' v \qquad [7.34]$$

$$m = t^{m} v \qquad [7.35]$$

where the ts refer to transition probabilities for unemployed, nonactives with and without benefits, and migrants. In the case study, it was assumed that there was no exogenous change in the number of vacancies. Because the number of vacancies being filled will equal the change in the number of jobs, the following will be true:

$$\Delta u = t^u (I - T)^{-1} \Delta e \qquad [7.36]$$

$$\Delta n = (t^n)' (I - T)^{-1} \Delta e \qquad [7.37]$$

$$\Delta r = (t^r)' (I - T)^{-1} \Delta e \qquad [7.38]$$

$$m = t^m (I - T)^{-1} \Delta e \qquad [7.39]$$

The vacancy chain model is tied to the input-output system through the income-consumption model. The change in consumption of employed people is:

$$\Delta c^{\ell} = (c^q)' w \Delta e + \Delta c_I' \qquad [7.40]$$

where c^q is the consumption vector by industry, w is the average labor income per industry, and the last term is the change in consumption that derives from changes in the exogenous incomes of workers. In the application to the northern provinces of the Netherlands, it was assumed that the latter term was zero. In addition, the part of labor income accruing to migrants has to be deducted from equation 7.40 to avoid double counting (as these migrants have already been included as part of the exogenous final demand). Hence, equation 7.40 becomes:

$$\Delta c^{\ell} = (c^q)' w (\Delta e - m) \qquad [7.41]$$

The decrease in consumption expenditures of the unemployed is derived in equation 7.42; the 0.80 coefficient is the product of the assumption that their unemployment benefits are approximately 90 percent of their former labor incomes and that they fill vacancies with incomes that are 90 percent of the corresponding sectoral average:

$$\Delta c^u = -0.80 (c^{qu})' w \Delta u + \Delta c_u' \qquad [7.42]$$

where c^{qu} is the consumption vector of coefficients for the unemployed and $\Delta c_u'$ the consumption changes from exogenous income changes of the unemployed (also assumed to be zero in the case study). The final decrease in consumption that will occur is associated with those who move from the status of nonactive with benefits to employment:

$$\Delta c^n = -12{,}227\ \Delta n + \Delta c_n' \qquad [7.43]$$

where $\Delta c_n'$ is the consumption from exogenous income (again assumed to be zero); and the value –12,227 is the average nonactive benefits, assumed to apply to all nonactives who get a job. The solution of the integrated model is shown in Figure 7.6. The model solution works in the following way; the exogenously induced effects (inmigrant expenditures) create industry demands that, through the operation of the interindustry multiplier, creates jobs. The vacancy chain model operates to fill these jobs and thereby to create income. These consumption-induced expenditures again create demands on industry and so the process continues until convergence. From the empirical model, the authors were able to show that from the 6700 families who entered the northern regions, 5000 persons gained employment in the region; the effect of this employment creation was the further creation of 2421 additional person-years of employment.

Linear Programming Input-Output Links

Two major limitations have curtailed the more widespread use of input-output models. The first relates to the problem of coefficient stability over time and the second to the problem of the absence of limitations on capacity in the regional economy. A number of authors have provided some creative ways of dealing with these problems, through the linkage of the input-output model with either a linear program or the respecification of the input-output model per se as a linear program. In the latter category, the work of Bargur (1969) and Mathur (1972) should be noted; Bargur developed a formulation in which a linear programming form of the input-output model was used to examine the effect of water capacity problems on output in Californian industries. Mathur's model examined interregional investment allocation—very much in the spirit of some of Ghosh's (1973) work; however, the MORSE model (Lundqvist 1981) is probably the most advanced because it employs a dynamic formulation. It was used for a variety of

Dimensions of the symbols:

	28	28	1	1	1	Endogenous variables:		Exogenous and lagged endogenous variables:
28	$(I-A)$	0	$-\underline{q}^1$	$-\underline{q}^u$	$-\underline{q}^n$	$\underline{\Delta x}$		$\underline{\Delta f^{ex}}$
28	$-\hat{1}$	$(I-T)$	$\underline{0}$	$\underline{0}$	$\underline{0}$	$\underline{V}$		$\underline{v}^{ex} - \hat{p}^1 \; \underline{e}_{t-1}$
1	$-(\underline{c}^q)' \hat{w} \hat{1}$	$\underline{0}'$	1	0	0	Δc^1	=	$\Delta c_1^{ex} - (\underline{c}^q)' \; \hat{w}\hat{p}^1 \; \underline{e}_{t-1}$
1	$\underline{0}'$	$0.80\,(\underline{c}^{qu})' \hat{w}\hat{t}^u$	0	1	0	Δc^u		Δc_u^{ex}
1	$\underline{0}'$	$12{,}227\,(\underline{t}^n)'$	0	0	1	Δc^n		Δc_n^{ex}

Figure 7.6 Integrated I-O Labor Model for the Dutch Northern Provinces

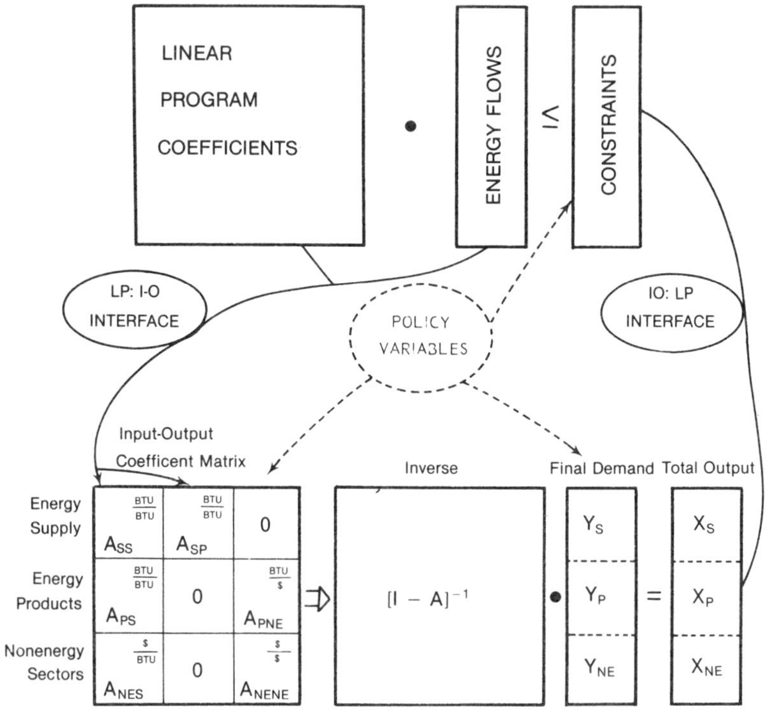

Figure 7.7 Generalized Solution for the Combined Input-Output Linear Programming Model

impact analyses—for example, the substitution of nuclear power in conjunction with increased oil prices and the case of increased regional self-sufficiency.

A similar concern for energy issues prompted the application of a linked input-output and linear programming model to part of a six-state region in the U.S. (Page, Gilmore, and Hewings 1981). The ORBES model structure is shown in Figure 7.7; note that the A matrix is divided into three major components: energy supplies, energy products (nontradeable items such as space heating, air conditioning), and nonenergy sectors. The *numeraire* varies by submatrix; essentially, final demands generate demand for energy products. These products are supplied from the set of available energy supplies through a linear programming model that seeks to minimize the cost of production subject to energy conversion, energy supply, capacity, and pollution constraints. Changes in relative prices of energy supplies are thus incorporated in the linear programming model and then, in turn, into the input coefficients (A_{ss} and A_{sp}) matrices. Thus, some degree of endogenous price substitution is incorporated into the system—if only in a limited number of sectors. Liew and Liew (1984) have approached this problem using production frontiers from which price frontiers are derived; regional technical coefficients, trade coefficients, and modal choice are determined by price and cost variables.

Further Extensions

The further set of linkages involving an input-output model centers on the role of the input-output model as a module of a larger, integrated set of models. Two examples will be described briefly here; the Isard and Anselin (1982) model is the precursor of an ambitious attempt to link national and regional econometric, demographic, investment, industrial location, input-output, and commodity flow models. Although many of the components have been developed, the linkages have not been made, and hence the empirical implementation may reveal some substantial problems as well as providing some new insights into the operation of the space economy. ORANI (Dixon, Parmeter, Sutton, and Vincent 1982) is a disaggregated model of the Australian economy, built in the tradition of the multisectoral models associated with Johansen (1960). These general equilibrium models have proved popular at the national level (see Adelman and Robinson 1980 for a Korean version) but ORANI is one of the few national models that has a regional component. To use Bolton's (1980) distinction, the model may be characterized as "top-down" because the Australia-wide data are apportioned to the six constituent states. The model for the regional disaggregation derived its rationale from the work of Leontief and Strout (1963); national level commodities created by an exogenous shock are allocated to the regions; a set of commodity-balance equations are used to solve for the production of regional outputs of local goods. Although the allocation rule for national commodities (tradeables in the social accounting nomenclature) ensures that the regional aggregates sum to the national totals, consistency is not assured for nontradeables. A separate set of equations ensures that the national and regional sums balance. The ORANI model enables a link to be made between international trade and regional economic activity, a feature that has not been common in many regional models (see Hewings 1982 for a review). Cavalieri, Martellato, and Snickars (1983) have provided a strong regional-interregional link; their model of Tuscany is cast in a two-region framework but contains the possibility of substitution between interregional and international trade. The reason for this is the fact that foreign exports drive about 25 percent of regional product.

Two final sets of links should be noted; the first integrates input-output and econometric models in the context of state/regional econometric models (L'Esperance 1980; Conway 1979). The final set of links focuses on input-output environmental models (Isard et al. 1971; Cumberland and Korback 1973; Miernyk and Sears 1974; Rose 1983). Cumberland (1966) provided a link between input-output, environmental analysis, and project appraisal in the context of a development project in Maryland.

Summary

Of all the developments in regional input-output analysis currently being pursued, the efforts in the direction of linked models offer some of the most exciting possibilities. The increasing sophistication of the modern economy has necessitated a concomitant increase in the range of modeling required to address current policy issues. By linking input-output with other models, the utility of the whole analytical system of models is often enhanced substantially. Over the next decade, we will probably continue to witness important new developments; some of these are described in the final chapter.

8. FUTURE DIRECTIONS AND RESEARCH NEEDS

The developments reviewed in Chapter 7 present many continuing challenges for research in input-output modeling. In addition to the efforts being directed toward linking regional input-output models with other models of the subnational economy, there are several avenues of inquiry concerning the input-output model itself currently being pursued. These will be reviewed briefly.

Relaxing the Assumptions

Several of the assumptions involved in the use of input-output models have tended to restrict their usefulness. The ORBES model described earlier provides an example of some of the potential benefits to be gained from the articulation of flexible input coefficients. Lahiri (1976) has been experimenting with the notion of scale-dependent coefficients; in this case, it would be assumed that the value of an input coefficient would depend on the level of demand for the product in the industry using the input. The Hudson and Jorgenson (1974) translogarithmic production function, in which relative prices essentially determine the production structure, provides another alternative, more flexible view. Rose (1984) provides a review of more than a dozen approaches to the issue of changing technical coefficients.

Gerking (1976, 1979a, 1979b) in a number of papers has challenged the notion of the input-output model being developed without error, in contradistinction to the major developments in econometric modeling. To this end, both Jackson (1983) and West (1983) have been pursuing ways in which properties associated with the distribution of firms making up a sector might be carried forward to the estimation of multipliers and other elements of the input-output model. In this fashion, it might be possible to specify upper and lower bounds to affect analyses in much the same way that one can reveal the confidence interval around an estimate in an econometric model.

Dynamic Modeling

With very few exceptions, almost all regional input-output models have been constructed for a single point in time. Any projections made with these models have been of the comparative static kind in which future estimates of the coefficients have been developed from a number of sources (especially national projections). Leontief (1970) proposed the formulation of a dynamic inverse matrix and Bruno et al. (1970) have developed a rather complex dynamic input-output model. However, at the regional level, there have been few contributions to this part of the input-output literature. Miernyk (1970) has implemented the most sophisticated regional dynamic input-output model; in this formulation, the investment demands are made endogenous. Two types of investment are specified: replacement investment (in which case, the stock of investment goods is not increased) and expansion investment. Because the use of investment functions requires some information on capacity levels, the issue of implementation quickly becomes very complex.

New Accounting Systems

The commodity-industry framework, in which a distinction is made between commodities and industries, has now become the accepted system in which national input-output models are developed. There have been few parallel attempts to produce similar accounts at the regional level. The advantage of the commodity-industry formulation is that it enables the analyst to avoid the assumption that industries produce only one commodity or that any given commodity can be produced in only one industry. A rectangular system of accounts has been developed so that greater commodity than industry detail may be revealed (see Jensen and Hewings 1985 for a review); this would facilitate linking the model with interregional commodity flow models and with models involving the consumption sector.

Some alternative formulations have been suggested by Beyers (1976) and Hewings (1983) to accommodate increasing interest in the role of large, multinational firms on regional economies. The input-output tables would be divided further into transactions involving multiplant firms and those involving local or single-plant operations. In this fashion, some insights into the degree of commodity ownership or control of production may be able to be gleaned from such a set of tables. In Sweden, where superior firm-specific data are available, Eliasson (1978) has been able to produce a micro-to-macro model of the economy. The standard macro framework of the kind described in this book serves as an upper level that sends signals to the micro (firm) level model. At this level, firms face uncertainty and attempt to produce a level of output on the basis of information received from the macro model about prices, wage costs, expected demand and, of course, the level of demand that existed in an earlier time period. The individual decision making firm is then "aggregated" back into the national model, adjustments made, and the whole process continues to a new iteration stage.

Use in Policy Analysis

Although regional input-output models have gained considerable use in impact analysis of the kind described in Chapter 4, the applications involving very strong links with policy formulation have tended to lag behind. Some early promise about the ability to link regional input-output models with some of the rich theoretical developments implicit in growth center theory appear not to have borne fruit. Attempts to identify key sectors along the lines suggested by Hirschman and Rasmussen (see Hewings 1982 for a review) have suffered from substantial inconsistencies. The area of industrial complex analysis has occasionally attracted the attention of input-output researchers; Czamanski and Czamanski (1977), Roepke, Adams, and Wiseman (1974), and most recently O'hUallachain (1984) have attempted to identify spatial clustering of industries at the regional level. However, there has been little attempt to use this information in the articulation of a space-time theory of regional development. Jensen and Hewings (1984) have been trying to understand the way in which the input-output structures of regional economies evolve through time. In what way does the economy become more complex and do the interrelationships among industrial sectors develop in some generalizable fashion? With so little empirical information on which to conduct such research, it may be many years before answers to these questions can be obtained.

Conclusions

The increasing attention being devoted to regional problems in developing economies has served to spur interest once again in analytical modeling at this scale. The development of more sophisticated nonsurvey and semisurvey techniques under conditions of limited information would appear to hold considerable potential in these countries. At the same time, the increasing ability being developed in linkage mechanisms of input-output with other regional models has provided a further stimulus to the continued development of regional input-output analysis. However, it is unlikely that input-output models will be developed independently at the regional level on a scale that occurred in the 1950s and 1960s in the U.S. The advantages of linked models would now appear to provide enough motivation to ensure their continued development at the expense of the isolated single-purpose model.

NOTES

1. Note that $(1 - a)^{-1}$ is equal to $[(1 + r/r)]$; because $a = 0.4$, $(1 - a)^{-1}$ is equal to 1.66. Hence we have a direct relationship between the economic base multiplier and the basic-nonbasic ratio.

2. In other frameworks, a distinction is made between commodities and industries. This avoids many of the problems of classification articulated here (see Jensen and Hewings 1985 for a review).

3. In Chapter 5, some discussion about survey and nonsurvey methods will be provided.

4. Investment goods are those products that are usually regarded as depreciable items, such as a machine, that the purchaser will depreciate over some finite time span greater than one year.

5. The entries in cells (1,1), (2,2), and (5,5) represent transactions among firms *within* an industry. Because our sectors are very highly aggregated, it is possible that some inter-

mediate, unfinished goods will be traded with firms that are classified as part of the same sector. In input-output tables with very large numbers of sectors the diagonal elements tend to be very small.

6. No distinction is made here between replacement and expansion capital along the lines used in the West Virginia study (Miernyk et al., 1970).

7. This follows from the fact that as size of the region increases, any one industry will lose its dominant position and hence the value of its location quotient will fall toward one.

8. In an n-region case, the corresponding expression to equation 6.3 would be $x_1 = B_{11}Y_1 = B_{12}Y_2 \ldots\ldots\ldots\ldots\ldots B_{1n}Y_n$.

REFERENCES

Adelman, I., and Robinson, S. 1978. *Income distribution policy in developing countries.* Palo Alto, CA: Stanford University Press.

Airov, J. 1967. Fiscal policy theory in an interregional economy: General interregional multipliers and their application. *Papers of the Regional Science Association* 19: 83-108.

Bacharach, M. 1970. *Bi-proportional matrices and input-output change.* Cambridge: Cambridge University Press.

Bargur, J. 1969. *A dynamic interregional input-output programming model of the California and Western States Water Economy.* Berkeley: University of California.

Batey, P.W.J., and Madden, M. 1983. The modelling of demographic-economic change within the context of regional decline: Analytical procedures and empirical results. *Socio-Economic Planning Sciences* 17: 315-328.

Batten, D. F., and Andersson, A. 1983. Towards a hierarchical system of models to coordinate national and regional economic developments. *Papers of the Regional Science Association* 51: 21-40.

Bell, C., Hazell, P., and Slade, R. 1982. *Project Evaluation in Regional Perspective.* Baltimore: Johns Hopkins University Press.

Beyers, W. B. 1976. Empirical identification of key sectors: some further evidence. *Environment and Planning A* 8: 231-236.

Blackwell, J. 1978. Disaggregation of the household sector in regional input-output analysis: Some models specifying previous residence of worker. *Regional Studies* 12: 367-377.

Blair, P. D., and Miller, R. E. 1983. Spatial aggregation of multiregional input-output models. *Environment and Planning A* 15: 187-206.

Bolton, R. 1980. Multiregion models: An introduction to a symposium. *Journal of Regional Science* 20: 131-142.

Brown, A. J. 1967. The "Green Paper" on the development areas. *National Institute Economic Review* 40: 26-33.

Bruno, M. et al. 1970. Dynamic input-output, trade and development. In *Applications of Input-Output Analysis*, ed. A. Carter and A. Brody. Amsterdam: Elsevier-North Holland.

Bullard, C. W., and Sebald, A. V. 1977. Effect of parametric uncertainty and technical change in input-output models. *Review of Economics and Statistics* 59: 75-81.

Bulmer-Thomas, V. 1982. *Input-output analysis in developing countries.* New York: John Wiley.

Burford, R. L., and Katz, J. L. 1977. Regional input-output multipliers without a full input-output table. *Annals of Regional Science* 11: 21-36.

Cavalieri, A., Martellato, D., and Snickars, F. 1983. A model system for policy impact analysis in the Tuscany region. *Papers of the Regional Science Association* 52: 105-124.

Chenery, H., and Clark, P. 1959. *Interindustry Economics.* New York: John Wiley.

Conway, R. S. 1979. The simulation properties of a regional interindustry econometric model. *Papers of the Regional Science Association* 43: 45-57.

Cumberland, J. H. 1966. A regional interindustry model for the analysis of development objectives. *Papers of the Regional Science Association* 17: 65-94.

———and Korbach, R. J. 1973. A regional interindustry environmental model. *Papers of the Regional Science Association* 30: 61-75.

Czamanski, D. Z. and Czamanski, S. 1977. Industrial complexes: Their typology, structure and relation to regional development. *Papers of the Regional Science Association* 38.

Dixon, P. B., Parmeter, B. R., Sutton, J., and Vincent, D. P. 1982. *ORANI: A multi-sectoral model of the Australian economy.* Amsterdam: Elsevier-North Holland.

Eliasson, G. 1978. *A micro-to-macro model of the Swedish economy.* Stockholm: Almqvist and Wiksell.

Erlander, S. 1980. *Optimal spatial interaction and the gravity model.* Berlin: Springer-Verlag.

Folmer, H., and Oosterhaven, J. 1984. *Modelling an interregional labour market projection incorporating the social security sector.* Paper presented at the Second Congress of Arts and Science, Rotterdam.

Gerking, S. D. 1976. Reconciling "rows only" and "columns only" coefficients in an input-output model. *International Regional Science Review* 1: 30-46.

———1979a. Reconciling reconciliation procedures in regional input-outut analysis. *International Regional Science Review* 4: 23-36.

———1979b. Reply to "Reconciliation procedures in regional input-output analysis." *International Regional Science Review* 4: 38-40.

———and Isserman, A. M. 1981. Bifurcation and the time pattern of impacts in the Economic Base Model. *Journal of Regional Science* 21: 451-467.

Ghosh, A. 1973. *Programming and interregional input-output analysis: An application to the problem of industrial location in India.* Cambridge: Cambridge University Press.

Golladay, F., and Haveman, R. 1976. Regional and distributional effects of a negative income tax. *American Economic Review* 66: 629-641.

Gordon, P., and Ledent, J. 1981. Towards an interregional demoeconomic model. *Journal of Regional Science* 21: 79-87.

Greytak, D. 1970. Regional impact of interregional trade in input-output analysis. *Papers of the Regional Science Association* 25: 203-217.

Haynes, K., and Fotheringham, A. S. 1984. *Gravity and spatial interaction models.* Beverly Hills, CA: Sage.

Herman, B., Mennes, L.B.M., and Waardenberg, J. G. 1969. Some exercises with a simple model for world development planning. In *Towards Balanced International Growth,* ed. H. C. Bos. New York: John Wiley.

Hewings, G.J.D. 1977. *Regional Industrial Analysis and Development.* London: Methuen.

———1982. Regional and interregional interdependencies: Alternative accounting systems. *Environment and Planning A* 14: 1587-1600.

———and Romanos, M. C. 1981. Simulating less developed regional economies under conditions of limited information. *Geographical Analysis* 13: 373-390.

Hoffman, R. B., and Kent, J. N. 1976. Design for commodity-by-industry interregional input-output models. In *Advances in Input-Output Analysis*, eds. K. R. Polenske and J. V. Skolka, pp. 251-261. Cambridge: Ballinger.
Hudson, E. A., and Jorgenson, D. 1974. U.S. energy policy and economic growth, 1975-2000. *Bell Journal of Economics and Management Science* 5: 461-514.
Isard, W. 1953. Some empirical results and problems of regional input-output analysis. In *Studies in the Structure of the American Economy*, ed. W. W. Leontief, pp. 116-184. Cambridge: Harvard University Press.
———1969. Some notes on the linkages of the ecologic and economic systems. *Papers of the Regional Science Association* 22: 85-96.
———and Anselin, L. 1982. Integration of multiregional models for policy analysis. *Environment and Planning A* 14: 359-376.
Isard, W., Choguill, C. L., Kissin, J., Seyfarth, R. H., and Tatlock, R. 1971. *Ecologic-economic analysis for regional planning*. New York: Free Press.
Isserman, A. M. 1980. Estimating export activity in a regional economy: A theoretical and empirical analysis of alternative methods. *International Regional Science Review* 5: 155-184.
Jackson, R. W. 1983. *A distributional approach to the modelling and simulation of industrial systems*. Ph.D. dissertation, University of Illinois, Urbana-Champaign.
Jensen, R. C. 1980. The concept of accuracy in regional input-output models. *International Regional Science Review* 5: 139-154.
———and Hewings, G.J.D. 1985. *Holistic measures of regional interindustry systems.* Urbana: University of Illinois.
Jensen, R. C., and West, G. R. 1980. The effect of relative coefficient size on input-output multipliers. *Environment and Planning A* 12: 659-670.
Johansen, L. 1960. *A multi-sectoral study of economic growth*. Amsterdam: Elsevier-North Holland.
Karunaratne, N. D. 1976. Quantification of sectoral development prospects in Papua, New Guinea using Tinbergen and Rasmussen criteria. *Developing Economies* 14: 280-305.
Kim, T. J., Boyce, D. E., and Hewings, G.J.D. 1983. Combined input-output and commodity flow models for interregional development planning: Insights from a Korean application. *Geographical Analysis* 15: 330-342.
King, L. J. 1984. *Central Place Theory*. Beverly Hills, CA: Sage.
Kuyvenhoven, A. 1978. *Planning with the semi-input-output method.* Leiden: Martinus Nijhoff.
Lahiri, S. 1976. Input-output analysis with scale-dependent coefficients. *Econometrica* 44: 947-961.
Ledent, J. 1978. Regional multiplier analysis: A demometric approach. *Environment and Planning* A 10: 537-560.
———and Gordon, P. 1981. A framework for modelling interregional population distribution and economic growth. *International Regional Science Review* 6: 85-90.
Liew, C. K., and Liew, C. J. 1984. Multi-modal, multi-output, multiregional variable input-output model. *Regional Science and Urban Economics* 14: 265-281.
Leontief, W. W. 1953. Interregional theory. In *Studies in the Structure of the American Economy*, ed. W. W. Leontief, pp. 93-115. Cambridge, MA: Harvard University Press.
———1966. *Input-output economics.* New York: John Wiley.
———and Strout, A. 1963. Multiregional input-output analysis. In *Structural interdependence and economic development*, ed. T. Barna, pp. 119-149. London: Macmillan.
L'Esperance, W. 1980. Conjoining an Ohio input-output model with an econometric model of Ohio. *Regional Science Perspectives* 7: 54-77.

Lundqvist, L. 1981. Applications of a dynamic multiregional input-output model of the Swedish economy. *Papers of the Regional Science Association* 47: 77-95.

McMenamin, D. G., and Haring, J. E. 1974. An appraisal of nonsurvey techniques for estimating regional input-output models. *Journal of Regional Science* 14: 191-205.

Mathur, P. N. 1972. Multiregional analysis in a dynamic input-output framework. In *Input-out techniques*, eds. A. P. Carter and A. Brody. Amsterdam: Elsevier-North Holland.

Mathur, V. J., and Rosen, H. S. 1974. Regional employment multiplier: A new approach. *Land Economics* 50: 93-96.

Miernyk, W. H. 1965. *The elements of input-output analysis.* New York: Random House.

———1976. Comments on recent developments in regional input-output analysis. *International Regional Science Review* 1: 47-55.

———1979. Comment on "Reconciling reconciliation procedures in regional input-output analysis." *International Regional Science Review* 4: 36-38.

———1982. *Regional analysis and regional policy.* New York: Oelgeschlager, Gunn and Hain.

———and Sears, J. T. 1974. *Air pollution abatement and regional economic development.* Lexington, MA: D. C. Heath.

Miernyk, W. H. et al. 1970. *Simulating Regional Economic Development.* Lexington, MA: D. C. Heath.

Miller, R. E. 1966. Interregional feedback effects in input-output models: Some preliminary results. *Papers of the Regional Science Association* 17: 105-125.

———1969. Interregional feedbacks in input-output models: Some experimental results. *Western Economic Journal* 7: 57-70.

———and Blair, P. D. 1981. Spatial aggregation in interregional input-output models. *Papers of the Regional Science Association* 48: 149-164.

———1985. *Input-output analysis: Foundations and extensions.* Englewood Cliffs, NJ: Prentice-Hall.

Miyazawa, K. 1976. *Input-output analysis and the structure of income distribution.* Berlin: Springer-Verlag.

Moses, L. 1955. The stability of interregional trading patterns and input-output analysis. *American Economic Review* 45: 803-832.

Mulligan, G., and Gibson, L. J. 1984. Regression estimates of economic base multipliers for small communities. *Economic Geography* 60: 225-237.

Nijkamp, P., and Spronk, J. 1981. *Multiple criteria analysis: Operational methods.* Aldershot: Gower.

O'hUallachain, B. 1984. The identification of industrial complexes. *Annals of the Association of American Geographers* 74: 420-436.

Oosterhaven, J. 1981. *Interregional input-output analysis and Dutch regional policy problems.* Aldershot: Gower.

Page, W. P., Gilmore, D., and Hewings, G.J.D. 1981. *An energy and fuel demand model for the Ohio River Basin Energy Study Region.* Urbana: University of Illinois.

Polenske, K. R. 1970. An empirical test of interregional input-output models: Estimation of 1963 Japanese production. *American Economic Review* 60: 76-82.

———1972. The implementation of a multiregional input-output model for the United States. In *Contributions to input-output analysis*, eds. A. P. Carter and A. Brody, pp. 171-189. Amsterdam: Elsevier-North Holland.

———1976. Multiregional interactions between energy and transportation. In *Advances in Input-Output Analysis*, eds. K. R. Polenske and J. V. Skolka, pp. 433-460. Cambridge: Ballinger.

Pyatt, G., and Roe, A. R. 1977. *Social accounting for development planning*. Cambridge: Harvard University Press.

Pyatt, G., and Round, J. I. 1983. Improving the macroeconomic data base: A SAM for Malaysia, 1970. *World Bank Staff Working Paper* 646.

Richardson, H. W. 1972. *Input-output and regional economics*. London: Weidenfeld and Nicolson.

Roepke, H., Adams, D., and Wiseman, R. 1974. A new approach to the identification of industrial complexes using input-output data. *Journal of Regional Science* 14: 15-29.

Romanoff, E. 1974. The economic base model: A very special case of input-output analysis. *Journal of Regional Science* 14: 121-130.

Rose, A. 1983. Modeling the macroeconomic impact of air pollution abatement. *Journal of Regional Science* 23: 441-459.

———1984. Technological change and input-output analysis. *Socio-Economic Planning Sciences* 18: 305-318.

———Nakayama, B., and Stevens, B. 1982. Modern energy regional development and income distribution. *Journal of Environmental Economics and Management* 9: 149-164.

Round, J. I. 1972. Regional input-output models in the United Kingdom: A reappraisal of some techniques. *Regional Studies* 6: 1-9.

———1978a. An interregional input-output approach to the evaluation of nonsurvey methods. *Journal of Regional Science* 18: 179-194.

———1978b. On estimating trade flows in interregional input-output models. *Regional Science and Urban Economics* 8: 289-302.

———1979. Compensating feedback effects in interregional input-output models. *Journal of Regional Science* 19: 145-155.

———1983. Nonsurvey techniques: A critical review of the theory and the evidence. *International Regional Science Review* 8: 189-212.

———1984. Decomposing multipliers for economic systems involving regional and world trade. Paper presented at the British Section Regional Science Association, Canterbury, Kent.

Schinnar, A. P. 1976. A multidimensional accounting model for demographic and economic planning interactions. *Environment and Planning A* 8: 455-475.

———1977. An eco-demographic accounting type multiplier analysis of Hungary. *Environment and Planning A* 9: 373-384.

Smith, P., and Morrison, W. I. 1974. *Simulating the urban economy: Experiments with nonsurvey input-output techniques*. London: Pion.

Steele, D. B. 1969. Regional multipliers in Great Britain. *Oxford Economic Papers* 21: 268-292.

Stevens, B. H., and Trainer, G. A. 1976. *The generation of errors in regional input-output impact models*. Regional Science Research Institute Working Paper A1-76, Amherst, Massachusetts.

Stevens, B. H. et al. 1983. A new technique for the construction of non-survey regional input-output models and comparisons with two survey-based models. *International Regional Science Review* 8: 271-286.

Stone, R. 1961. Social accounts at the regional level. In *Regional Economic Planning*, eds. W. Isard and J. H. Cumberland, pp. 263-269. Paris: Organisation for European Economic Cooperation.

———1963. *A Programme for Growth*. Cambridge: Cambridge University Press.

Tiebout, C. M. 1969. An empirical regional input-output projection model: The State of Washington, 1980. *Review of Economics and Statistics* 51: 334-340.

Tinbergen, J. 1966. Some refinements of the semi-input-output method. *Pakistan Development Review* 6: 243-247.

Treyz, G., Friedlander, A. F., and Stevens, B. H. 1980. The employment sector of a regional policy simulation model. *Review of Economics and Statistics* 62: 63-73.

Van Dijk, J., Oosterhaven, J., and Stoffelsma, R. J. 1984. Regional impacts of inmigration: An input-output/vacancy-chain approach. Paper presented at the British Section RSA Conference, Canterbury.

West, R. G. 1983. *Sensitivity and probability analysis in regional input-output models.* Ph.D. dissertation, University of Queensland, Australia.

Wilson, A. G. 1970. *Entropy in urban and regional modelling*. London: Pion.

ABOUT THE AUTHOR

GEOFFREY J. D. HEWINGS is Professor of Geography and Regional Science and of Urban and Regional Planning at the University of Illinois at Urbana-Champaign where he is also Head of the Department of Geography. Prior to his move to Illinois, he taught at the Universities of Kent at Canterbury, England, and Toronto, Canada. He received his Ph.D. (1969) and M.A. (1967) from the University of Washington, Seattle, and his B.A. (Hons.) (1969) from the University of Birmingham, England. His research interests lie in the application of urban and regional models under conditions of limited information and he has been associated with regional model development in England, Australia, the United States, Greece, and Korea. Previous monographs include *Regional Industrial Analysis and Development* and (co-editor) *Industrial Location and Regional Systems.* He has also served as Executive Secretary of the Regional Science Association (International) since 1978.